IVORY FORTRESS

IVORY FORTRESS

A PSYCHIATRIST LOOKS AT WEST POINT

Richard C. U'Ren, M.D.

THE BOBBS-MERRILL COMPANY, INC.

Indianapolis/New York

Published by the Bobbs-Merrill Company, Inc.
Indianapolis   New York

ISBN 0-672-51923-2
Library of Congress catalog card number 73-22666
Designed by A. Christopher Simon
MANUFACTURED IN THE UNITED STATES OF AMERICA

First printing

For

HERBERT E. BOWMAN

and

GEORGE SASLOW

ACKNOWLEDGMENTS

I would like to acknowledge my gratitude to Janette Barbour for her remarkably efficient typing and unfailing good spirits under pressure; to my wife Marjorie for her thoughtfulness and intelligent criticism; to Frank Conrad, who was the ideal friend and colleague while we were stationed at West Point together; to Diane Giddis, my editor at Bobbs-Merrill, whose suggestions, thoughts, and encouragement helped me to make this a much better book than it otherwise would have been; and—most of all—to the cadets and officers at West Point who were willing to talk openly and freely with me.

CONTENTS

PREFACE

I was chief of psychiatry at West Point from 1970 to 1972, while the Vietnam war was at its height. My assignment to the Academy gave me an unprecedented opportunity to observe the most prestigious institution of American military life. West Point is the heart of the army. Not only are West Pointers the elite of the army's officer corps, but the standard set at West Point is considered the ideal for the rest of the army.

Throughout my tour of military duty, I devoted a considerable amount of time to ferreting out answers to questions any outsider is sure to have about West Point. I relied most heavily on interviews and conversations with cadets and officers. The literature on West Point, unless one has a taste for boys' adventure stories, turned out to be almost useless. There is no end to the number of adulatory books about the Point, but I was struck by the absence of any work that describes Academy life realistically.

No one had to tell me that an establishment like West Point would be loath to stand inspection in anything but its best uniform. As I expected, I met resistance from the beginning. The first officer I asked simply refused to be interviewed. It was not until I assured him of complete anonymity that he finally relented. When at last we met, he began by saying, "These days one has to be careful. I'm still not exactly sure what I'm in for, but last night I decided someone has to tell the truth about West Point."

This book is a version of the "truth" as I saw it, based upon my own observations and many interviews with officers and cadets both inside and outside my office. I was able to learn a great deal about the Academy as I interviewed over one hundred cadets for psychiatric purposes, though of course that was not my principal aim in those meetings. I tried to make at least brief notes after almost every conversation I had with cadets and officers at social gatherings at West Point during my two-year stay: parties, dances, dinners, desserts, sports events, classes, lectures, seminars, and meetings. By the time I left the Academy, I had accumulated well over three hundred pages of notes from these conversations alone. Another important source of information consisted of publications from the Academy's Office of Institutional Research. I was able to study and take extensive notes from more than thirty of their documents while I was stationed at West Point. And, finally, I held in-depth interviews at length with cadets and officers outside my office. Each of the interviews lasted at least three, and sometimes as long as eight, hours. I tried to select cadets and officers with as diverse viewpoints as I could find. Fifteen cadets and eighteen officers, all but three Academy graduates, were included.

The aim of this book is to show Academy life as I saw it directly and as I saw it through the eyes of cadets and officers at West Point. My original intent was to write a kind of institutional travelog about the Academy. As I worked with the material I had collected, however, I became aware that it was leading me to certain conclusions. I began to write more of an analytical, critical study of the Academy than a simple exposition. My study has none of the characteristics of a rigorous scientific inquiry. I entertained no highly developed hypotheses around which to gather data while I was still on duty at the Academy. Although I made it a point to cover certain topics in my in-depth interviews, much of my information is anecdotal. Accordingly, the book is studded with anecdotes and stories that were related to me. I have sought to make West Point come alive as it came alive for me.

As readers will soon discover, the book is by no means confined to my psychiatric observations about the Academy. Instead, I have attempted to look at the entire Academy and assess how and whether it carries out its mission, where it succeeds and where it fails, and what kind of person it attracts, molds, and produces.

Throughout the writing of this book I was faced with two problems. The first was the problem of generalization. Did I have a representative sample of people? Did I have enough information to allow me to come to fair and reasonable conclusions? The number of people I talked with, large though it was, could not possibly embrace all the opinions, thoughts, and feelings of four thousand cadets and eight hundred officers. Of course, it would have been nice to interview all of them, but this was impossible; so I had to accept the fact that my material was necessarily incomplete and to make the most of it. Being a psychiatrist at an institution as illustrious as West Point may tempt one to make exaggerated claims on the basis of limited experience. I have tried to resist that. After all, my two years of military service were spent entirely at the Academy and obviously provided a limited experience with the vast world of the army, which would take more than a lifetime to know intimately.

The second problem was more complicated and focused around the West Point cadets and officers. Are West Pointers the way they are merely because the Academy attracts and selects certain young men who have already developed certain personality traits? Or are they that way because it alters them in some fundamental way as they go through four years there? The answer, I think, is both. If it does not exactly implant certain attributes of character, the Academy certainly magnifies certain traits, while it allows others to atrophy quietly. It is crucial to understand the kind of person West Point attracts, but I have no doubt that the Academy has enormous impact on developing young men. What that impact is and what its consequences and implications are for West Pointers, the Academy, the army, and the larger society I have tried to suggest in this book.

What authority do I have to tell the "truth" (in the officer's words) about West Point? There have been no military people in my background, no earlier contacts with military life. I was a civilian who was drafted for a short time as a doctor with the somewhat gratuitous rank of major. I can therefore lay no claim to being an expert in the realities of military life. I can, however, claim to be a trained observer of human behavior who spent two intensive years in a unique position at the Academy, who had access to information only rarely available to civilians, and who commands the perspective of the outsider if not always the knowledge of the professional soldier. The advantage of being an outsider may be especially great in a military establishment so intolerant of criticism that no regular officer on active duty is ever likely to divulge its secrets.

Virtually all the cadets and officers with whom I talked wish to remain anonymous. I have respected their wishes. I have cited references whenever possible. Getting information from the Academy was often difficult. Toward the end of my tour at West Point I was able to arrange a meeting with a sympathetic acquaintance in the Office of Institutional Research. I talked there with an older staff member who had been involved in many projects carried out in that office over the years. He admitted how frustrating it was to him that his research was never published outside the Academy: "Last week we presented some of our material to the Dean. He said, 'That's good work, but make sure none of it gets out.'"

This book is as accurate and current as I can make it. I know how fiercely military personnel will attack any inaccuracies in a work critical of their foremost institution. Though I left West Point in July 1972, I have talked at considerable length with several cadets and officers who are still there. I feel that my information is reasonably up-to-date. Several recent Academy graduates and West Point officers have proofread the manuscript, but I know that some errors are inevitable. These, I hope, will not seriously impair the validity of what I have to say about America's most famous military academy.

IVORY FORTRESS

1

THE WORLD OF
WEST POINT

The U.S. Military Academy occupies a commanding position on the west bank of the Hudson River fifty miles north of New York City. Many people, including those who apply for admission, think of it as a college. Its well-publicized football teams, its four-year curriculum, and its faintly academic appearance certainly suggest that. The observant visitor can quickly detect the true military nature of the place, however. The powerful stone buildings bear the names of famous generals— Grant, Lee, Washington, MacArthur, Eisenhower; the homes that flank Thayer Road may have been designed by Stanford White, but army colonels occupy them now; the library is guarded by cannon, and a statue of George Patton, binoculars in hand and a pistol on each hip, gazes upon it from across the road.

West Point is a world of its own, a separate reality. This shows in many ways—clothing, for example. Civilian student attire usually expresses a variety of taste. Not so West Point dress, which is strictly prescribed according to the season of the year and the occasion. Student clothes these days reveal very little about an individual's status in the civilian world. The Academy uniform indicates a young man's achievements, his cadet rank, and his class year at the Academy.

Vocabulary, too, is different from that of the civilian world. One could make a list of parallel words covering practically all aspects of living: "quarters" instead of "house"; "bunk" instead of "bed"; "mess hall" instead of "dining room"; "fatigues" instead of "work clothes." In the military world of West Point, civilian vocabulary becomes taboo. No one at the Academy would use the term "dining room" except humorously. Similarly, "regulations" at West Point carries the same meaning as "law" in civil life.

Even "student body"—the collective term for students at civilian colleges—has a different name at West Point: the "corps of cadets," which is organized in a decidedly noncivilian fashion. The superintendent, a lieutenant general, is the commanding officer at West Point. Two other general officers on post, both brigadier generals, are the commandant of cadets, who has responsibility for military training and athletics, and the dean of the academic board, who has responsibility for cadets' education. The corps is organized under the supervision of the commandant into a brigade with four regiments. Each regiment has nine companies, each of about one hundred and ten men. Regular army colonels supervise each of the regiments, while army captains or majors—"Tactical Officers"—supervise each of the companies.

The military world of West Point is unlike the civilian world for a good reason: it has a different purpose. Whatever its other roles are, the Academy's reason for being is to train men to fight and win battles. This is the end to which all the values and rules of West Point are subordinate.

The stated ideals of West Point are inscribed on the ubiquitous Academy crest: Duty, Honor, Country. Each ideal has its own specific meaning. "Duty" is the sense of obligation which leads one to do his best in every situation, a readiness to accept and loyally execute all assigned missions and a willingness to "live within the spirit of all regulations or directives regardless of origin." "Honor" means integrity, "complete integrity of both word and deed." The Cadet Honor Code— "A cadet does not lie, cheat, or steal"—is the guardian of the West Point sense of honor. "Country" means love of country. It is surprising how little the cadet handbook, *Bugle Notes,* has to say on country. As Gore Vidal recently pointed out, "It is curious that no one until recently seems to have made much of the ominous precedence that makes the nation the third loyalty of our military elite."[1]

"Duty" at West Point primarily means obedience, the cardinal military virtue. A profession dedicated to serving and displaying the armed might of the state must be organized into a "hierarchy of obedience." Orders come from the top; lower levels of the hierarchy must carry them out instantaneously and unquestioningly. Obedience is the first and central lesson cadets receive at West Point. Any man entrusted with power, with weapons of destruction, must be taught that they are never for his own use; he must always be held accountable to others. It is not too far-fetched to say that the entering cadet's first summer at the Academy is obedience training. He must obey literally hundreds of orders, commands, and regulations every day. Anything but instant response brings swift and immediate reprisal.

Appearance is an integral part of duty at West Point: A cadet should always "set the example in terms of soldier-like appearance and bearing." A cadet's appearance, in the words of one major, is "the outward sign of invisible grace." The military man's attitude toward appearance was neatly summed up by a 1915 West Point graduate, Dwight D. Eisenhower: "I deplore the beatnik dress, long unkempt hair, the dirty necks and fingernails now effected by a majority of our boys. If universal military training accomplished nothing more than to produce cleanliness and decent grooming, it might be worth the price tag. . . . To me a sloppy appearance has always indicated sloppy habits of mind."[2]

Appearance, "looking good," signifies pride ("Posture is pride," says a poster at West Point), obedience, discipline, and loyalty. A cadet's polished shoes and shining belt buckle show his willingness to place the arbitrary requirements of the group above his own needs.

The primacy of the group is an important theme at West Point. Success in any military activity requires the subordination of the will of the individual to the will of the group. To ensure the success of the military mission, soldiers must not only respond without hesitation, but work together as a team. Hence the heavy stress at West Point on uniformity and cooperation.

To be treated exactly like everyone else is probably the first experience of military life anywhere. The process is called "equalization" at West Point and is carried out "because of the diverse background of entering cadets and the need for absolute impartiality in the operation of the military personnel system."

A cadet said, "One advantage of the system, I think, is that everyone can come here—a farmer's son or a colonel's son—and do equally well. Colonel Borman's son [Colonel Borman was the first American astronaut to circle the moon] came here because he thought it was the only place he'd be treated like everyone else." While it would be too extreme to say that all cadets look alike, their politeness, neat uniforms, and military bearing give an impression of sameness. How successful West

Point is in obliterating differences arising from background became clear to me at a gathering I attended for several graduating cadets and their parents in June 1972. One cadet's father, from Virginia, came dressed in a Brooks Brothers suit, while another's, from Louisiana, wore a blue bowling shirt; yet it was impossible to determine which cadet belonged to which father. Equality under stress fosters group solidarity; the purpose of uniformity is to produce an instant, shared background, a common starting point in the interminable military race for rank and honor. The stress on conformity and uniformity is so heavy at West Point that to be different is to be in trouble.

"Cooperate and Graduate" is a common cadet saying. Because the multitude of tasks imposed upon them is so great, cadets learn to help each other early in their careers. In learning interdependence, they also lay the groundwork for intense loyalty to each other. A well-established but informal Academy tradition holds that a cadet who is academically superior to his classmates will help others rather than race ahead at his own speed. Every week during the academic year, for example, an upperclassman designated as the academic sergeant receives a "deficiency list" on each man in his company who is having trouble with academic work. This academic sergeant is then responsible for seeking out and assigning each of these men a tutor. "The academic sergeant makes sure everyone passes," said a cadet.

Yet at the same time the Academy promotes competition in a multitude of ways, for the "will to win" is a highly esteemed military virtue. The reward for superior competence on the battlefield may be life rather than death. Candidates selected for West Point have already proved themselves to be academically, socially, and athletically competitive. West Point intensifies this competitiveness. In first-year math classes, for instance, men are graded and ranked every day, six days a week. Throughout every cadet's four years at the Academy, an evaluation of virtually every activity in which he participates is fed into a complex formula which determines his class ranking in "the general order of merit." High standing is important, since it determines a cadet's choice of service branch, his first assignment in the army, and his order of promotion in later years. "Your order of merit follows you around forever," a cadet remarked. Intramural athletics and intercollegiate sports are means *par excellence* of encouraging competition. The best company teams in each regiment are feted at regular intervals throughout the year, and pictures of the winning teams are posted on the walls of the cadet gym.

The mission, a well-defined objective or goal, is paramount in the

4

army. Getting the job done always has priority over any other consideration. At West Point all activities are rigidly scheduled. This presumably teaches a man to use his time efficiently, to set priorities, to function under stress, and to make decisions rapidly. In combat, an officer must be able to act decisively and directly; hesitation is fatal. Time is money at West Point, an association not wasted on the authorities. Time given and time taken away are the primary reward and punishment used to shape desirable behavior.

Through its unyielding insistence on these values—obedience, group loyalty, cooperation, uniformity, competition, and the importance of the mission—West Point creates and maintains its own reality, a world as divorced from the "outside world" (a common military expression) as the Academy can possibly make it. Each cadet who attends West Point enters this world and, unless he leaves before his four years are up, never fully emerges from it. In its attempts to impose its reality on those it admits, West Point is undeniably successful.

But just as undeniably, West Point's success is also its failure. For its monastic system, devoted so single-mindedly to the development of fighting men, accounts for so many of the problems, dissatisfactions, weaknesses, and deficiencies of West Point training.

2

THE WEST POINT CANDIDATE

Who goes to the U.S. Military Academy? What kind of person chooses to commit himself to the world of West Point?

Vital statistics reveal that the successful West Point candidate is eighteen years old, weighs one hundred and sixty pounds, and is five feet, ten inches in height. When I first arrived at West Point, I expected to see a cadre of brawny football players. This was not the case; most West Pointers are not physically imposing. A West Point doctor, one of the few who had himself been a cadet, maintained that summer training had become more difficult for new cadets now than in the past because of a subtle shift in body-type within the corps. Whereas in the past the predominant body-type of cadets was mesomorphic (muscular), it was now, he thought, ectomorphic (thin and wiry). Mesomorphs, being stronger, have more endurance than ectomorphs. I do not know whether the Academy has statistics that would document such changes, but I do know that William Sheldon, a Columbia University psychologist who studied cadets in the 1950s, found them to be uniformly mesomorphic. In fact, the only group so consistently mesomorphic, he told me, were prisoners in the Oregon State Penitentiary.

In spite of their modest physical appearance, however, entering cadets have impressive athletic credentials. In the entering class of 1975, eighty-five percent had participated in varsity sports, seventy-seven percent had earned a varsity letter, and thirty-one percent had been captains of one or more of their high schools' athletic teams. Fifty percent of the entering class had played football, the most popular sport.

Activities and leadership credentials were equally impressive. Ninety-one percent of the class of 1975 belonged to a high school club, and forty-five percent were either president or vice-president of it. Sixty-seven percent were involved in clubs outside of school, and a third of entering cadets were elected president or vice-president of these. Five percent were president of their senior class, six percent were president of their student bodies, and nine percent were a senior class officer other than president.[1]

Academically, cadets are good—far better than the average entering college student—but not so good as students at the best colleges or perhaps even at the other two service academies, Annapolis and the Air Force Academy. Seventy-one percent of the entering class of 1975 at West Point ranked in the top twenty percent of their high school class, ninety percent in the top two-fifths. Eight percent of the 1975 class were either first or second in their graduating high school class, forty-two percent had completed a high school honors course, forty-nine percent belonged to the National Honor Society, and thirty-seven percent had rejected academic scholarships elsewhere. Compared with entering freshmen at colleges throughout the country who participated in the American College Testing Program, cadets scored in the ninety-second percentile in English and in the ninety-third percentile in math.*

On the Scholastic Aptitude Test, however, cadets ranked lower. In the 1975 class, the average score on the verbal section was 554 (the seventy-first percentile compared with high school seniors accepted at other colleges) and, in math, 622 (seventy-seventh percentile). First-rate civilian colleges attract more students who score over 700 on the college boards than does West Point. Forty-one percent of Harvard's 1972 class, nine percent of Vanderbilt's, forty-nine percent of Cal Tech's, and fourteen percent of Rensselaer Polytechnic Institute's scored above 700 on the college board exams in verbal ability; the comparable figures for the 1972 class at the Air Force Academy, Annapolis, and West Point were seven percent, six percent, and seven percent, respec-

---

* A percentile score tells what percentage of the group falls below that particular score. If an individual has a percentile score of ninety, that means that ninety percent of the group taking the exam received a lower score than he did.

tively. In mathematical ability, fifty percent of Harvard's 1972 class, twenty-two percent of Vanderbilt's, ninety-six percent of Cal Tech's, and forty-eight percent of Rensselaer Polytechnic Institute's scored over 700 on the college boards. At the Air Force Academy, Annapolis, and West Point, the comparable figures were thirty percent, twenty-five percent, and twenty-two percent. Scores from 1972 showed that West Point entrants do slightly less well on the college boards than students entering Annapolis or the Air Force Academy; thirteen percent of the class that year had verbal ability scores below 499, while only five and one-half percent of entering cadets at the Air Force Academy and six percent of midshipmen at the Naval Academy had verbal scores below that level. In the great middle range, however, entering students at the academies were more alike: forty-eight percent of West Point cadets, forty-nine percent of Air Force Academy students, and fifty-one percent of Naval Academy midshipmen had scores in verbal aptitude which fell in the 500–599 range.[2] Though forty-four percent of the entering West Point class of 1973 stood in the top tenth of their high school graduating class, the comparable figure at the Air Force Academy was fifty-three percent.[3]

Few of the cadets I knew had applied to top civilian colleges. It was rare to find someone who had thought seriously about Yale or Stanford; most of them had planned to go to state university or a junior college near their home.

Cadets come from all over the country. Although the greatest single number of cadets entering the Academy in 1971 were from New York State (134, almost ten percent of the class),[4] the northwestern and western states were somewhat overrepresented at West Point in earlier classes, while the midwestern states were underrepresented. Contrary to popular belief, the southern states are not overrepresented at the Academy.[5]

Compared to civilian colleges and universities, the Military Academy attracts fewer students from large cities and more from moderate-sized towns and cities of 50,000 or less. In fact, the percentage of cadets from large cities is only half the percentage of students from large cities in four-year colleges and one-third the percentage of students in technical institutions and private universities. Most entering cadets graduated from high school senior classes of about 400 students.

Surprisingly, and contrary to general opinion, alumni sons made up only five percent of the 1975 class—a figure smaller than many Ivy League schools. However, twenty-three percent had fathers who were career military men, and only seventeen percent of cadets in the entering class had fathers who had no military service whatsoever.[6] The big difference between cadets at West Point and freshmen at other colleges and universities is the number whose fathers are military careerists— anywhere from nineteen to twenty-five percent of each class. At four-year colleges, only 1.8 percent of boys come from military families, while the figure rises to seven percent at technical institutions.

Ninety-two percent of entering cadets grew up in homes in which both parents were present and forty-two percent of cadets were first-born. The average size of each cadet's family was five.[7]

Another surprising finding is that cadets' parents, typically, have more formal education than parents of students enrolled at a wide variety of four-year colleges and technical institutions, and as much as parents of students at private universities. In the 1972 class, thirty-nine percent of fathers of entering cadets had a high school diploma or less, twenty-five percent had earned a bachelor's degree, and sixteen percent had obtained a postgraduate degree. However, less than four percent of the class had fathers who had earned a medical, law, or Ph.D. degree. Fathers of cadets in the class of 1972 most commonly listed their occupations as businessmen (twenty-eight percent), military careerists (nineteen per-cent), or skilled workers (ten percent). Private university students listed their fathers' occupations as businessmen (forty-two percent), skilled workers (ten percent), or engineers (eight percent).

Most cadets come from families with an income of between ten thousand and fifteen thousand dollars a year; thirty-five percent come from families earning less than ten thousand dollars and only three percent from families earning thirty thousand dollars or more. In con-trast, thirty-one percent of students at private universities have parents whose yearly income is less than ten thousand dollars, but fifteen percent have parents who earn more than thirty thousand dollars.[8]

Information about their religious preferences indicates the middle- and lower-class origins of most cadets. Only about one-fifth of new cadets belong to denominations—Episcopal, Presbyterian, or Congre-gational—which attract high-status families. Most cadets belong to other Protestant denominations or to the Catholic Church. Indeed, the per-centage of Catholics (thirty-two percent) who enroll at West Point is higher than the percentage of Catholics in the American population (twenty-six percent). Several cadets who were raised as Catholics told me they had no trouble adjusting to the discipline of Academy life; their fathers were more strict than the Academy! The percentage of Jews at West Point—less than one percent—is lower than their percentage in the population at large (three percent). The largest minority group at

9

West Point is black (about three percent of the entering class of 1975), followed by Americans of Oriental origin (less than one percent).[9]

In their aspirations and involvements, entering cadets are more interested than freshmen at civilian institutions in keeping up-to-date with political affairs, becoming an authority in a special field, taking administrative responsibility for the work of others, becoming a community leader, and having an outstanding athletic career. They are less concerned than other freshmen with being successful in a business of their own, being well off financially, or making a contribution to science or to the visual or performing arts. Remarkably, seventy-seven percent of new cadets expect to obtain a master's degree or a doctorate during their military service. In contrast, fifty-eight percent of students in private universities expect to get a master's or a doctorate.[10]

Cadets are considerably less liberal than their civilian counterparts. A significantly greater proportion of cadets in the 1972 class, when compared to their civilian colleagues, felt that college officials were too lax in dealing with student protests, that student publications should be cleared by college officials, that college officials have been right to ban persons with extreme views from campus, and that college officials have the right to regulate student behavior off-campus. In general, cadets' views toward these matters are similar to those of freshmen at technical institutions.[11]

With regard to political views, one percent of cadets in the class of 1974 described themselves as "far right," thirty percent as conservative, thirty-seven percent as "middle-of-the-road," twenty-three percent as liberal, and one percent as "far left." (Eight percent did not answer the questionnaire.) But while civilians expect to become more liberal during their college years, West Point cadets expected to become less so. By graduation, three percent of cadets expected to be "far right," forty percent conservative, twenty-nine percent "middle-of-the-road," and eighteen percent liberal. None expected to be "far left." Cadets were also less inclined to liberalize divorce, legalize marijuana, abolish capital punishment, or concede the right to publish all scientific findings.[12]

In summary, a higher proportion of entering cadets received A's, were in the top ten percent of their high school graduating class, were elected president of their student organizations, and won a varsity letter than students who went to civilian colleges. One Academy study concluded that successful candidates for West Point were, as high school students, "more likely than other students to attend a religious service, discuss sports, work in a school political campaign, discuss their futures with their parents. Cadets were less likely than other students to vote in a student election, argue with their teachers in class, play a musical instrument, ask a teacher for advice after class, do extra reading for a course, work in a political campaign. . . . In general, the activity patterns of cadets are more like those of students at technical institutions."[13]

But what do these statistics boil down to? To a young man whom I shall call Michael D. Hines, an eighteen-year-old high school senior from a small town in eastern Pennsylvania who sought admission to West Point in the entering class of 1976. Polite and refreshingly candid, Hines was a good-looking young man whose grayish eyes and longish light brown hair went well with his checkered sport jacket, white shirt, and wide blue tie. He weighed 150 pounds and was 5'10".

Hines had lived his entire life in the same town where his father, now forty-seven, managed a local department store. His mother worked three days a week as a licensed practical nurse in the local hospital. The only other family member was a sister, Ruth, twenty-four, who had married after one and a half years of secretarial school. Michael was close to his father: "Dad is always busy, but he takes time to ask me how things are going. Until a couple of years ago, we'd pass the football or shoot baskets together. We always got along well. Dad complains that he's working too hard, but that's typical. He wouldn't know what to do with himself if he weren't. Dad has said several times that he wished he'd gone on with college instead of quitting after two years—but that seems to be his only regret." His father was drafted toward the end of the Second World War, spent two years in college on the G.I. Bill, then went to work.

Hines himself had always planned to attend college. "I've been hearing about that since the fifth grade! Grades have always been a big thing around the house. I have a cousin, Rick. Everyone in the family knows which of us gets the better grades each term. I don't think my sister went through the same thing—Mom and Dad didn't seem to get as upset if she'd come home with an occasional C. But they care what I do, Mom especially." Hines had responded to this pressure. "I had mostly A's and B's except for my freshman year English, when I got two C's. I've always had to work for my English grades." At the end of his junior year he ranked fifty-fourth out of 350 classmates. He took the college boards in November 1972 and scored 620 in Math, 525 in English. "I've always done better in math and science," he explained. "It comes easier." One teacher wrote of him: "His strengths are his drive, his willingness to meet goals and deadlines, and his good mind. I have students who are more imaginative, but few have Mike's reliability and perseverance."

Hines won varsity letters in football and baseball, though he did not make first-string varsity football as a halfback until his junior year. "My first year was pretty mediocre. But this year our team came in second in the conference, and I was even given honorable mention on the all-conference team." Because of his small size, he did not expect to play football at the Academy if he was accepted. "No one came around recruiting, either," he added.

"I've never had any trouble getting along with other guys," he said. His popularity was confirmed his senior year when he was elected vice-president of the class. "I'd been president of our church group, but senior vice-president is a bigger thing," he said. No one was happier about his election than his steady girl friend, Joan.

As far back as he can remember, Hines's interests have always been sports, carpentry, and science. His political and social attitudes are largely conservative and traditional. For example: "If we hadn't fought in Vietnam, I'm afraid the Communists would have taken over." He has had little experience with drinking and has not smoked marijuana. "It's around, and I've seen it a couple of times, but I've just never felt like trying it," he said. Hines's family are devout Lutherans.

Why did he apply to West Point? His favorite uncle, who was an officer in the Korean conflict, excited his interest in the military by his many stories about the war. Until he came to West Point for a visit, Hines had never met anyone who had been at the Academy. "But you hear of it all the time," he said. "I guess I really got interested in West Point my sophomore year in high school. My science teacher came here for a one-day visit.* When he came back, he told me West Point did things better than anyplace he'd ever seen. When I see him at school, he still asks me if I'm going to apply.

"I believe in America. Every country has to have a strong military, doesn't it? And when I looked into coming here, I found that it's free and that cadets actually get a salary. Dad said he would help send me to Penn State, but that's not West Point. I think I'd really like being an officer. There's no telling where a person might end up. Two presidents went to West Point," he added enthusiastically.

"Mom's really happy I've applied. Already she's told all her friends. Dad is quieter and keeps saying it's my decision, but you can tell he's pretty happy about it too."

At the time I met Mike Hines, I thought he had an excellent chance of

---

* Groups of selected high school counselors come to the Academy throughout each academic year from all over the country. Travel expenses, but not room and board, are paid by the government.

getting into the Academy. He was typical of the successful candidate: upwardly mobile, ambitious, competitive, hard working, of middle-class origin, conservative, relatively but not exceptionally bright, athletic, popular, and friendly.

Why *do* young men come to the Academy? Although almost impossible to rank precisely, the most common reason is in order to get what they think is a good education. Candidates, when they visited West Point, often said something like, "It's the best education in the country" or "It's the most complete education I can get." And one, perhaps closer to the truth than he knew, wrote, "A West Point education makes you more advanced and ready to except [sic] the world." The desire for a military career was mentioned next most frequently. I suspect this embraces such diverse motives as patriotism ("I want to serve my country") and a romantic concept of the army ("Military life will be exacting, exciting, and dangerous").

Many cadets are attracted by the character-development training that West Point provides. Statements such as "I think West Point will make a better person out of me" or "It will give me the discipline I need" were common. These responses reflect, of course, the upward mobility of candidates who are anxious to improve themselves. Young men also apply to West Point because of its prestige and status. West Point, it seems, has always been able to project a favorable image of itself to the public. For many applicants and their parents, the West Point name is as prestigious as Harvard's.

The class that entered the Academy in 1970 agreed that the attitude of many (forty-seven percent) of their fellow high school students toward the military was negative; only twenty-eight percent of their closest friends shared this negative view, however. Nine out of ten entering cadets were aware of bad publicity the army received from the NCO club scandal, the Green Beret case, and the My Lai incident, but eighty-seven percent said these incidents had little or no effect on their decision to attend West Point. It would seem that neither the army nor West Point had acquired a bad image for most cadets because of Vietnam. High school students really unhappy about the war would never have applied to West Point in the first place. But, on the other hand, the Academy might well have held a greater attraction for other students who wished to escape the draft and serve as officers instead of enlisted men and, at the same time, to obtain an education at government expense.

Naturally, the way a young man thinks the Academy stacks up against civilian schools is itself a powerful motivator. Entering members of the

1974 class thought that West Point would offer them a greater sense of belonging, more opportunities for making lasting friends, a greater chance of realizing their "fullest potential," and more opportunity to participate in intramural athletics. In only two ways did they think a civilian college would have an advantage: it would offer them more time to study and more opportunity to make their own decisions on "important matters."[14]

A free education and a desire to please family and friends were also important considerations for applying to and entering West Point. In a time when schools as well known as West Point are so expensive, a government-sponsored education which also pays a salary of three hundred dollars a month can be irresistibly attractive to parents and candidates alike. Parents are overwhelmingly encouraging and supportive when their son considers the Academy. I was surprised how often they themselves had filled out their son's application. In fact, most adults—relatives, high school teachers, coaches, and neighbors—are very enthusiastic about a young man's desire to attend the Academy. Only close friends or girl friends are sometimes apt to try to persuade a successful candidate to turn down his appointment to West Point.

A wish for security, leadership training, the opportunity to participate in sports, and a desire for independence round out the reasons candidates apply to West Point. More than one candidate for the class of 1974 said, "I would like to come to West Point because it's a place where there are no riots." But this need for security also has other dimensions. If a cadet enrolls at West Point at the age of eighteen, he has virtually sealed his vocational fate for at least nine more years: four at the Academy as a cadet and five more as a commissioned officer. His career is, in fact, assured if he behaves himself. I suppose this need for security can be seen most clearly when cadets are threatened with the loss of it. When forced to resign for one reason or another, the first question cadets seemed to ask was, "Where do I go from here? I'm not ready to make a decision about what I want to do."

At the same time, though, many cadets are attracted to the Academy because it offers them independence. If they attend the Academy, they will not be indebted to their parents for a college education. I think they have a misconception about an officer as an independent person, a person who makes decisions forcefully and decisively. They rarely seem to consider that an officer takes orders from a superior.

In fact, most cadets have only a vague notion of what West Point really is or how difficult it can be. In response to a questionnaire they were given a few days after they arrived at West Point in July 1970, twenty-five percent of the class said they had no idea at all or a completely inaccurate idea of what military life would be like; another thirty-one percent said they had a pretty good idea, but it was inaccurate in several significant ways. Indeed, only forty-four percent of the class saw West Point correctly—primarily as a military school. Thirty-three percent came to West Point in spite of the Academy's disciplinary system, thirty-six percent came in spite of the five-year obligatory service after graduation, and forty-seven percent came despite what they knew about West Point's "traditional system for training new cadets."[15]

Since a high proportion did not know much about either West Point or the life of an officer, it is not surprising that a large percentage— up to twenty percent—of the first-year class leaves before the year is over or that many graduates, thirty to thirty-five percent, leave the army after their obligatory tour of service is over.[16]

Many cadets, then, come to West Point for reasons other than to make a military career, e.g.; a free education, prestige, and status. But although most of them come to West Point to receive an education, the majority of them do not see themselves as "serious" students.[17]

West Point has traditionally been able to attract good students. Whether they will be able to continue this tradition is another matter. While I was stationed at West Point, there were persistent rumors— denied by the admissions office, of course—that the quality of cadets was dropping each year and that in 1968 two hundred or so vacancies in the first-year class were left unfilled. I do not know whether or not this is true. I do know that the size of the corps of cadets has been increased from approximately 2,500 men in 1962 to 4,000 in 1974. This increase, probably in combination with other factors, such as a tarnished military image and the five-year military obligation after graduation from West Point (in the past it was only three or four years), has diluted the quality of recent classes.

Twelve percent of entering cadets in the class of 1967 were either valedictorians or salutatorians of their graduating high school class; the figure had dropped to eight percent for the 1975 class. In 1963, seventy-six percent of new cadets were in the top fifth of their high school class; the figure was seventy percent for the entering class of 1975. In 1963, fifteen percent of the entering class were either student body or senior class presidents; in 1970 the comparable figure was ten percent. A difference in parents' education, an indicator of social and economic status, also appeared between 1972 and 1974. Forty-one percent of fathers and twenty-five percent of mothers of sons in the 1972 class had a college or a postgraduate degree. By contrast, only thirty-six percent of fathers and nineteen percent of mothers of the 1974 class had the same academic credentials.[18]

Although the leadership potential and aptitude for physical activity was as high for the class of 1975, as for classes which preceded it, the Military Academy has watched applicants' college board scores decline steadily since 1967. Although this has been true for colleges all over the country—in 1965 the average verbal score for all students taking the exam was 471, the average math score 496, as compared to respective average scores of 454 and 487 in 1971—at the Military Academy the decline has been more spectacular. In 1965 the average verbal and math scores for entering cadets were 578 and 655; in 1971 they were 554 and 622.[19] In the same six-year period, the average national verbal and math scores declined by seventeen and eleven points, respectively, while at West Point the scores dropped twenty-four and thirty-three points.

Though uncertain about the exact reasons for this decline, West Point is of course worried about it. Some officers, however, have taken a positive view. Said one lieutenant colonel, "I've heard several people say that we might do better with this group—cadets with lower college board scores—than with a better academic group." What he meant was that less bright cadets might be easier to train and retain in the service than brighter ones.

Laurence Radway of Dartmouth College, among others, has raised the possibility that the impact of the service academies on our society may well be determined more by what kind of students they attract in the first place than by the experiences students have at the academies. If this is true, the decline in academic and leadership credentials among entering cadets in the past several years is obviously a matter of some concern.

3

BEAST BARRACKS

Beast Barracks, officially called "New Cadet Barracks," is a two-month period of intensive basic training each summer, beginning in the first week of July. It is worth close attention since it shows so clearly what West Point, the real West Point behind the glamor and tradition, is all about. Beast Barracks is an impressive, if sometimes brutal, demonstration of the Academy's first priority: military training. There is no pretense of education; academic work does not begin until September.

"Beast," as cadets call it, is named for the new cadets who enter West Point each July. "They're completely wild, untrained young men who need discipline," said an Academy graduate. The senior cadet in charge of this summer training detail is known as "The King of Beasts." But this term for summer training undoubtedly has more primitive associations; in one West African initiation ceremony, a demon or beast in the form of a large and terrifying mask is used to frighten young initiates into a state of insensate terror, representing death, from which they are rescued only by members of their tribal religious society, representing rebirth.

Though indoctrination and training take place throughout a cadet's four years at West Point, they are most intense during the first year and especially during the first summer. Beast Barracks is indoctrination at its rawest. Perhaps indoctrination is too crude and incomplete a word to describe all that happens at West Point during the summer. Beast Barracks is really West Point's rite of passage, the ritual in which young men are transformed into soldiers, the ordeal by which a civilian is initiated into the military world. New cadets learn from the beginning the cardinal military virtues of obedience and loyalty. Unquestioning, instant, automatic obedience to superiors is the principal lesson of Beast Barracks, as it is the unending lesson of West Point.

Initiation rites everywhere share the same purposes: to teach young men a new role, to instruct them in control and self-restraint, and to eliminate those who are unfit or unsuitable. Lionel Tiger and Robin Fox note:

Initiation procedures vary, but there are some standard elements. The initiates are separated from the women and kept in seclusion. They are hazed and humiliated by their elders. They undergo ordeals of endurance and tests of many skills. They are often physically mutilated. . . . They are compelled to learn masses of arcane wisdom, as well as the proper conduct of ritual and the proper cherishing of myths and traditions of the group. They often indulge in homosexual practices. Finally, they are sometimes ritually slain and brought back to life as "men." Isolation, endurance, testing, mutilation, and the internalization of tribal wisdom in some form or other crop up in society after society as each in its peculiar way copes with the problems of tapping and controlling the ambivalent energy* of the initiate.[1]

Mutilation and homosexual practices are not part of the West Point rites, but there is no doubt that the similarities between Beast Barracks and initiation procedures the world over are much more striking than their differences.

In order to convert civilians into soldiers, West Point applies enormously powerful emotional and physical stresses to eighteen-year-olds. Beast Barracks is the most shocking experience of a new cadet's life. Most of them, never having been away from home or parents before, find themselves engulfed in a totally alien culture characterized by constant hazing and incessant physical activity which demands that they relinquish their former individuality and freedom completely. Investi-

---

* The "ambivalent energy" to which Tiger and Fox refer is the sexual and aggressive activity at puberty and during adolescence associated with markedly increased levels of the male hormone, testosterone. "Suddenly [young men] are the most energetic and aggressive members of the community. But they have not had time to learn to cope with this infusion of novel and stormy information from the ancient centers of the brain."

gation of army recruits during basic training has shown that certain biochemical changes associated with stress are higher than at any other time in their military career, even including combat.[2] The same is no doubt true for new cadets at West Point.

Induction into the military world at West Point is abrupt for new arrivals. Many parents bring their sons to the Academy. New cadets take leave of their parents and their civilian existence as they pass through one-way turnstiles at the football stadium, where the initial stages of processing take place. I shall never forget the looks of pride, yet also anxiety and uncertainty, on the faces of parents as they stood rather forlornly in small groups outside the stadium following their sons' departure. It was obvious they had no clearer idea of what their sons were in for than their sons did. Once inside the stadium, every new cadet is physically and mentally—there is no other word for it—assaulted. His civilian clothing and possessions are confiscated; he is issued regulation military apparel. He is addressed by upperclassmen only as "mister," yet must himself learn the names of those who harass him and must address them only as "sir." His hair is cut short,* a procedure which emphasizes to the cadet that he no longer has control over his own appearance; he now belongs to the military. Every new cadet is weighed, measured, and photographed in a jockstrap. He is forced to run from place to place and told exactly how to sit and how to stand. In response to a question or statement from an upperclassman, he is allowed only one of four responses: "Yes, sir," "No, sir," "Sir, I do not understand," or "No excuse, sir." Every new cadet is viewed not as an individual but as a faceless member of a large group badly in need of military training. The whole experience can be devastating. As one cadet said, "It was so bad there was almost no one who didn't think of resigning at some time during Beast. Most cadets have forgotten how degrading that time is, but I haven't." Cadets find it hard, and sometimes painful, to describe or remember their first summer at West Point. I suspect this poor memory is symptomatic of the overwhelming stress they feel. It is a simple inhibiting response of the brain to stresses it cannot deal with by any other means, and is a well-known phenomenon in wartime.

An atmosphere of fear, anxiety, and tension pervades the lives of new

---

* The repugnance that men, especially military men, feel toward long hair is intriguing. The obvious interpretation is that long hair is associated with women; cutting a new cadet's hair removes a threat which some men evidently feel when a woman joins their group. Hair is Biblically associated with strength and power. Delilah cut Samson's hair to render him helpless and weak.

cadets. The system is almost entirely mistake-oriented. Upperclassmen ruthlessly subdue any show of independence on the part of new cadets; errors are caught and corrected immediately. "There is no place in the military profession for an excuse of failure," says one West Point handbook. A former cadet remembered, "You lived with the sense you'd inevitably get caught at something." It was of course impossible to avoid mistakes. Another cadet recalled, "The first time I got chewed out for anything was when a—I think he was a major—came by and asked me what the serial number of my rifle was. I said, 'forty-nine thirty-seven,' when I should have said 'four-nine-three-seven.' My squad leader, when he heard about it later, came by and chewed me out, too."

A third cadet, when asked about Beast Barracks a month after he had successfully completed it in 1971, said, "The first two weeks were the worst. I don't think I'd have come through it if I'd known how we'd be treated. I cried every night. I can't really say how I made it. Being able to see the humor of it might have been the most important thing. Absurd things that would have never seemed funny at other times were hysterically funny then. Once the table commandant told me I could take a banana back to barracks to eat. As I was walking across the area, an upperclassman stopped me and asked me why I was carrying a banana. When I told him I had been given permission, he said, 'Well, carry that banana like a rifle then.' We had Inspection Banana. I had to peel it on one side for him. He stuck his finger in it and said, 'Dirty bore.' "

Another cadet recalled, "Meals were the worst time during Beast Barracks. If you got hassled during a meal, it meant you had no break from hassling all day." New cadets learned self-sacrifice and cooperation by agreeing before meals which of them should eat first. The first man to take a bite was sure of being harassed the rest of the meal by upperclassmen, but this at least allowed the other cadets at the table to eat in relative peace. This harassment, presumably in the service of training, sometimes seemed unending. One new cadet, after a short stay in the hospital during the third week of Beast Barracks, was not allowed to eat a full meal for several days after he returned to his barracks. He had already eaten too much in hospital, his table commandant informed him.

The Academy, however, recognizes that it is important for new cadets to eat. Knowing that zealous upperclassmen may prevent this, the commander of Beast Barracks, an army colonel, may order the food on a particular table to be weighed before and after a meal. If too much is left, the upperclassman in charge will get "slugged" (punished). The operation does not always guarantee that the new cadet will be properly fed, however; his share can always be eaten by someone else.

An upperclassman once accidentally spilled gravy on a new cadet's pants during lunch. Since the cadet had no chance to get back to his barracks and change his pants, he was criticized for his sloppiness the rest of the day. Just before regular inspection, a new cadet may be ordered to put all his clothes hangers in his mattress cover and bring it to an upperclassman for inspection. This, of course, means the new cadet must undo his bed, remove his mattress cover, take all his clothes off the hangers, have them inspected, replace his mattress cover, and remake his bed—all during the time he should be preparing for inspection. Also, just before inspection, a new cadet may be ordered to appear in all his uniforms before an upperclassman. He then has to run up and down the stairs in his dress uniform, his tropical worsteds, his gym clothes, and so on.

Of course, all orders are given in language that is insulting in tone if not in content. If a new cadet fails to respond promptly to an upperclassman's question, the upperclassman will shout, "Irp, Irp," in his face. This is a crude demand for "Immediate Response, please." Upperclassmen are discouraged by their officers from using personally insulting language with new cadets, but it happens nevertheless.

New cadets look depressed and frazzled during Beast Barracks. Their letters home during this time are epistles of discouragement. The Academy does forewarn parents of this and suggests how they can best deal with it. But some parents, alarmed, demand that the Academy investigate their son's complaints. Others take more drastic measures. One mother, whose son did not initially want to leave the Academy but who nevertheless wrote the usual disheartened letters home, came in person to West Point and demanded his release. When refused, she took up a lonely vigil in the bleachers across from the barracks and remained there three days and nights until her son was allowed to resign.

It is no wonder that cadets and parents alike are distressed, though most, I suspect, would find it difficult to say exactly why. In addition to severe harassment, the new cadet experiences the loss of his family, his civilian world, and even his previous identity when he enters West Point. These losses may be tantamount to small deaths and no doubt mark the beginning of a long process which serves to numb an individual psychologically, to blunt his sensibilities, and to impress upon him the arbitrary power of authority.

To these ends, every week, every day, every minute of a new cadet's time is scheduled. There is little opportunity to be alone and almost no chance to find privacy. Out of 682 scheduled hours of activity during Beast Barracks, the largest share—157 hours—is devoted to tactical training: marches, use of the bayonet and rifle, squad tactics, and so on.

Ninety-nine hours are consumed by general military training, which includes ceremonials, cadet drills, inspection, and manual of arms. Eighty hours are taken up by administrative tasks—testing in preparation for the academic year, picking up required clothing, processing and reception, and so forth. Sixty-eight hours are reserved for physical education. The other 278 hours are devoted to cadet-oriented training, chapel, and other activities. New cadets count only 52 of the total 682 hours as completely their own.[3]

Every new cadet is always working under pressure of time; he is forever being stretched further than he would have imagined. Reveille comes at 5:40 in the morning. He must be in ranks by six o'clock and at breakfast a few minutes later. After breakfast he returns to barracks, helps clean them, gets his room in order, and prepares for the day's activities. This is followed by physical conditioning exercises on one of the athletic fields for forty minutes or so, after which he returns to barracks, changes uniforms, and gets ready for another formation. He may then be marched to the cadet store to pick up uniforms or to Thayer Hall, where he will receive instruction in military justice or human relations. He might then go to another class in Thayer Hall or to the parade ground for a lecture on the honor code. By eleven o'clock, the new cadet might return to his room in order to prepare for an inspection or to memorize such required material as the mission of the U.S. Military Academy, cadet songs, the American soldier's code of conduct, Schofield's definition of discipline (which begins, "The discipline which makes the soldiers of a free country reliable in battle is not to be gained by harsh or tyrannical treatment"), the number of lights in Cullum Hall (340), the number of gallons of water in Lusk Reservoir (seventy-eight million), the definition of leather ("the fresh skin of an animal, cleaned and divested of all fur, fat, and other extraneous matters . . ."), and the answer to "What do plebes rank?" ("Sir, the Superintendent's dog, the Commandant's cat, the waiters in the mess hall . . . and all the Admirals in the whole damn Navy").

Dinner formation takes place at 12:10 P.M., followed by a drill period at one o'clock. In the first week of drill, new cadets learn to march in squads of ten men. As training goes on, they learn to march in platoons of fifty men, then in companies of 150–160 men. Toward the end of Beast Barracks, squad drill is dropped and cadets practice close-order marching only in platoons or companies. The last of eighteen drill sessions ends with competition at the platoon level. As a matter of fact, all activities during Beast Barracks become progressively more complicated and difficult as time goes by. In order to develop "cardiorespiratory endurance and muscular strength," for example, cadets begin one of their conditioning programs by running six minutes and walking two at a pace of eight to nine miles per hour for fifteen minutes. At the end of the fourth week of training, they must run nine minutes for every two minutes of walking, for a total of thirty minutes. At the end, cadets are expected to run thirty minutes, without walking, at the pace of nine miles per hour. Similarly, marches become progressively more difficult, starting with two three-hour marches, proceeding to two four-hour marches, and ending with two marches of four and a half and seven and a half hours each.

After an hour of drill ending at two o'clock, the new cadet will go to Thayer Hall for another hour of class in military etiquette or the use of the M16 rifle. After marching back to his barracks and changing clothes, he must then go to mass athletics every day from 3:30 to 5:30. The schedule is planned so that every cadet spends a week playing one sport—soccer, football, tennis—before moving on to another.

Between 5:30 and 6:00, the new cadet must return to barracks, shower, and prepare for supper, which begins shortly after six. When the meal is over at 6:40, he must return to his rooms for the evening, where he usually spends his time taking care of his equipment or memorizing. Sleeping is of course strictly forbidden. On some evenings, however, cadets must take tests, prepare for a march the next day, or attend a "shot formation" at the hospital. The day officially ends at 10 P.M., but many cadets are forced to stay up later shining shoes, polishing brass, or finishing other tasks which they failed to complete during the day. "It seems like you never get all the work done in time," said one new cadet. If they are caught staying up late, however, they face harassment and demerits and must go to bed without completing their unfinished business.

If one cadet is especially far behind, his roommates will help him catch up. He, of course, will do the same for his roommates if they are in trouble. This willingness to help each other is the foundation of the durable personal loyalties cadets form with each other. Roommates must also learn to cooperate to avoid punishment. If an inspecting officer finds something amiss, like dirt under a bed, all three roommates will hear about it.

New cadets' mealtime behavior is strictly regulated by upperclassmen in accordance with Academy tradition. They must sit bolt upright, cut their food into morsels no larger than a pat of butter, and recite their poop—a large body of information consisting of Academy lore and "The Days" (the menu for the meals of the day, the movies for the week, days until Christmas leave, graduation, etc.)—on demand. Such rigors—designed to teach a young man how to function under stress, to hold his temper under control, and, indeed, to test his motivation for a military career—hamper eating, of course. Sometimes cadets can lose as much as thirty pounds during Beast Barracks, and few are fat to start with.

The techniques West Point uses on its new cadets during Beast Barracks—isolation, fatigue, tension, and the use of vicious language—are similar to the techniques used in thought reform. There are gentler words and terms for this remarkable process, such as "conversion" or "attitude change," but the intent is the same: to make men more vulnerable to new ideas, attitudes, and behavior. Naturally, the content of indoctrination is different at West Point than it is, say, in Communist China, but the procedures are strikingly alike. R. L. Walker, in *China under Communism*, describes how the Communist Party trained individuals to serve as a "transmission belt" between the Party and the masses. This special training lasted six to nine months. The first phase of the conversion process was called "the phase of physical control" and lasted two months. Throughout this period of training, six factors were present:

First, the training takes place in a special area or camp, which almost completely severs all ties with the trainees' families and former friends and facilitates the breakup of old behavior patterns. A second constant factor is fatigue. Students are subjected to a schedule which maintains physical and mental fatigue throughout the training period. There is no opportunity for relaxation or reflection; they are occupied with memorizing great amounts of theoretical material and are expected to employ the new terminology with facility. Coupled with the fatigue is a third constant: tension. . . . Uncertainty is a fourth factor throughout the process. . . . Trainees who conspicuously fail to comprehend the camp pattern in the first few weeks disappear overnight, and there is usually a well-sown rumor concerning their fate. . . . A fifth constant factor is the use of vicious language. . . . The final factor is the seriousness attached to the whole process. Humor is forbidden.[4]

Uncertainty and bewilderment about both the purpose of their rigorous ordeal and their day-to-day existence is always a problem for new cadets. Not only do they have but the vaguest idea of what will happen to them, they are also uncertain of the consequences of misbehavior. What would the Academy really do if a cadet went absent without leave? Would he (they wonder) be dismissed from the Academy, be punished, or be liable for a court martial? No one can, or will, tell them. The creation of uncertainty is intentional; it forces the cadet to realize that he is no longer his own man. It also places great pressure on cadets to conform. Safety lies only in following rules.

This distressing uncertainty is heightened by the unpreparedness almost all of the cadets have for Academy life. The public image of West Point, associated with successful football teams, American presidents, and the Long Gray Line, tends to obscure for applicants the quintessential fact that West Point is the military. The Academy itself often neglects to announce its purposes as clearly as it might. All of the service academies, of course, have to look attractive if they hope to attract top candidates. This is not easy at a time when the military is so vastly unpopular. West Point recruiters tend to present the Academy more as a college than as the military training institute it is. "I didn't really know what I was getting into when I applied here," said a West Pointer who graduated in 1972. "I remember my mother telling me I'd have the cultural advantage of being near New York City." There is also a bland, abstract quality about military language which cloaks certain realities of military life. The mission of the Military Academy is "to instruct and train the corps of cadets so that each graduate shall have the qualities and attributes essential to his progressive and continuing development throughout a career as an officer in the Regular Army." The mission of an army in war is to defeat the enemy. While there are ways to go about this which do not necessarily involve slaughter, one of the most distasteful yet fundamental tasks a soldier has to be trained for is to kill other people. A West Point officer said, "What attracts people to West Point is the image of someone like Robert E. Lee sitting nobly astride his horse Charger gazing over a fog-enshrouded valley. What they don't see is the carnage beneath the fog." The fundamental goal of the Academy is to train young men in the management of violence, but the stark realities of extreme situations—war and death—seem far removed from West Point.

The events of Beast Barracks are very deliberately planned. Beast Barracks, though much more strenuous and concentrated, is like basic training everywhere in the army. The first two weeks are the most difficult for new cadets because the upperclassmen are so zealous about carrying out their new mission as trainers and because new cadets are unprepared for the harsh realities of military life. By the third and fourth weeks, however, some of the enormous pressure is eased, at least temporarily. For one or two days at a time, new cadets escape the confines of training barracks to go on short marches during the day. In the field, soldiers are always less formal with each other; West Point upperclass-men let up on their weary charges and allow them to recoup. "There was no hassling at meals—you could finally eat by yourself—and it was wel-come because it was a break in the routine," said one cadet.

Until this time, every new cadet has received very little in return for his sacrifice. He has been personally degraded and forced to live in a way totally alien to what he has always known. He has been reduced to an obedient but inexperienced automaton. But as the first month of training comes to an end, several events take place which allow new cadets to enter what psychiatrist Peter Bourne has called "the period of attainment." (Bourne, in a study of new recruits undergoing basic training at Ft. Dix, conceptualized the training period as divided into four stages: 1) the period of environmental shock, which results from the recruit's sudden immersion in the military environment; 2) the period of engagement, which lasts until the third or fourth week and is associated with the personal mortification described above and the rudi-ments of basic training; 3) the period of attainment; and 4) the period of termination, occurring about a week before the end of basic training, a phase "marked by great euphoria, a feeling of immense confidence, and open expression of a sense of invincibility."[5])

The most important event of the attainment phase occurs when cadets are taken to the rifle range for several days in order to qualify with their weapons. Bourne writes, "Rifle scoring seems to mark the turning point for the trainee: It is the first time that the army gives him credit for an acquired skill, particularly one on which such a high premium is placed in the military." As a cadet said of the experience, "The first day was spent sighting the rifles, and the second day we fired long and short range at pop-up silhouettes. It was the highlight of the summer, especially since I qualified 'expert.' "

From this point on, new cadets become more and more proficient in their soldierly skills, which serves as an antidote to their faltering self-esteem and severe demoralization. By the end of the third week of training, parents are allowed to visit their sons again. The Academy evidently feels that cadets' military identity is secure enough by then to risk exposure to the core of their former civilian existence, their parents. By the end of the third week, also, new cadets are permitted to go to "the Boodlers," the cadet snack shop, where, for the first time since their arrival at West Point, they can purchase candy, cookies, and soft drinks.

The Boodlers (a great event in this time of deprivation), parents' weekend, and several days on the rifle range serve to buttress cadets for an event they face with considerable apprehension: the replacement of the upperclass detail. In the middle of Beast Barracks a fresh group of upperclassmen take charge of new cadet training. Persistent rumors about the harshness of the second detail, usually initiated by upperclassmen themselves, heighten the anxiety of new cadets. In the tense, unpredictable atmosphere of Beast Barracks, they feel at a terrible disadvantage in the face of these reinforcements. But in fact the second detail is no harder than the first.

The final test and culmination of Beast Barracks takes place near the end of August. Cadets are ordered on a long hike to Lake Frederick, where they bivouac for several days and then return. Those who complete the hike know they have withstood the worst of Beast Barracks. Upon their return they are escorted the last mile by an army band, cheered by spectators along their line of march through post, and paraded en masse before the superintendent's quarters. Besides a test of endurance, this ceremony is an initiation rite in miniature: boys are sent into the wilderness and come back "reborn" as young men, a transformation acknowledged by everyone on base.

Summer training is now at an end. Until 1973, one trial remained for new cadets which marred the elation that came with the end of Beast Barracks: "Reorganization Week," called "reorgy" by cadets. During the last week of August, new cadets were placed in the companies to which they would belong during their entire four years at the Academy. But the entire corps also returned to West Point that week. "Reorgy" gave upperclassmen not only the opportunity to see how well the new cadets had been trained, but also the chance to displace their own frustrations on the prospective plebes. "When I wanted to get my rocks off," said a junior, "I gave it to the new cadets." This harassment was no different in kind than new cadets had received throughout Beast Barracks; there was simply more of it. Newly returned upperclassmen called upon new cadets more often to recite their poop at the mess hall table, criticized them more harshly for mistakes, and carried out inspection with more than their usual zeal.

At the end of Beast Barracks in 1973, traditional "reorgy" was altered. Deciding that another week of ill treatment on top of Beast Barracks was more than flesh could bear, the Academy required the upperclassmen to return and settle in for several days before the new cadets were marched back from Lake Frederick. "The upperclassmen didn't see the new cadets when they first came back this year," said a senior. "That made it a lot easier for new cadets. Upperclassmen settled down, their anger at coming back settled; everyone was beginning to relax by the time the new cadets returned."

The capstone of the entire summer now comes at the end of Re-organization Week with an acceptance parade. New cadets officially become members of the corps of cadets at this ceremony and are called "plebes" rather than "new cadets" for the rest of the year. They are now presumably ready to face academic work and the first year at West Point.

That Beast Barracks accomplishes its goals is beyond doubt. Anyone who compares the ragged, unhappy boys of July 1 with the disciplined, more confident cadets of August 31 can see this impressive difference. Not everyone survives Beast Barracks, however: ten percent or more of the entering class may drop out before the summer is over, and in the first two weeks of Beast Barracks in 1973, fifty of approximately 1,300 cadets resigned. Beast Barracks may also have unintended conse-quences which do not show up till later. Thirty percent of the class of 1971 said that Beast Barracks was the unhappy experience that decided them against an army career.[6] One can imagine the profound disillu-sionment cadets must feel when they face the realities of military life for the first time. What a shock Beast Barracks must have been for the young, idealistic high school graduate who said to his father just before he left for West Point, "There's lots of things that need changing, and you have to be in a place where you can do something."

Beast Barracks is an unforgettable experience for every cadet. But whatever its long-range effect might be, it does pay important dividends for those who remain. Beast Barracks can markedly enhance a cadet's self-esteem and sense of personal strength. He knows he has passed the most rigorous and demanding test that West Point can give. He feels accepted by a group of men for whose approval he has sacrificed his civilian life. Perhaps most important, every new cadet who survives Beast Barracks begins to forge the powerful personal loyalties that bind him to fellow West Pointers. Confronted with seemingly unbear-able stress and isolated from his parents and his friends, every new cadet must literally depend upon his classmates in order to survive. Every psychiatrist knows that powerful emotions, pleasant or unpleasant, often strengthen bonds between people, that group solidarity is enhanced when members of a group go through intense emotional experiences together, and that anxiety and uncertainty heighten a person's reliance on those around him. So do generations of West Point officers, who have designed Beast Barracks to cause in young men the emotional turmoil which forms the basis of personal bonds and group identity so vital to the army's elite.

The stated mission of Beast Barracks is to "indoctrinate, motivate, and equip" new cadets with the physical and military skills they will need in order to join the corps of cadets. The major lessons of Beast Barracks, however, are obedience to superiors and loyalty to each other. These are, in fact, the major lessons of West Point which will be reinforced endlessly over the next four years.

A CADET'S FOUR YEARS

Beast Barracks is the first and perhaps most drastic lesson in obedience and conformity that cadets receive at West Point. But Beast is also a portent, in exaggerated form, of the next four years. "Things don't really get better," said one dejected sophomore; "you just get used to them." Since cooperation and discipline are esteemed so highly in the military, individualism and self-reliance—the old civilian virtues —must be ruthlessly expunged. The group with its mission, not the individual with his desires, comes first at West Point.

To subdue the healthy animal spirits of 4,000 young men is no easy task, of course, but West Point has had 172 years of practice. To have any chance of bending each cadet to its will, the Academy must have total control of his life.

The daily existence of generations of West Pointers has been regulated by the Academy's regulations manual, the notorious "Blue Book." Revised and shortened in 1973, the new edition has not yet achieved the notoriety of its predecessor, but its regulations are as pervasive as the old, in spirit if not in detail. It contains six chapters, the first of which is a general section that defines the purpose of the regulations ("to establish the standards, authorizations and privileges for cadets during the academic year"), what a cadet should do in the absence of written or oral instructions ("exercise his own good judgment and common sense"), the concept of duty, the meaning of seniority within the corps ("Seniority is determined first by cadet rank and thereafter by alphabetical order among cadets of equal rank within each class").

The second chapter spells out standards of behavior: whom cadets should salute, deportment in the cadet mess, courtesy at athletic events (including the injunction, "Standing on the sidelines is inconsiderate of the majority of spectators whose view of the field is thereby obscured, and also is distracting to the players"), how finances should be handled, courtesy at lectures, and the Academy's policies on hazing ("prohibited"), alcoholic beverages, political activity ("unauthorized"), and giving out information to the public.

The third chapter—and the second longest in the book—covers uniforms and appearances. "Soldierly bearing as well as the appearance of the military uniform and the manner in which it is worn are significant indicators of individual pride and unit discipline, morale, and esprit. For this reason, cadets are expected to exercise correct posture and to maintain their uniforms and personal appearance in a meticulous manner at all times." The emphasis in this chapter is on when, where, and how cadets should wear each of their fourteen different uniforms.

The fourth chapter specifies the regulations for barracks and quarters: what to do with family visitors who unexpectedly show up while a cadet is serving a "punishment tour," what electrical appliances are allowed in rooms ("beginning with the second term of their fourth class year, and provided they are table model types, cadets may have one radio and one set of stereo components per room. All other electrical appliances are unauthorized"), how windows should be open ("only from the top"), the purpose of the orderly room ("primarily a place of business"), use of the elevators ("cadets of the upper three classes may not ride elevators between 0715 and 1830 on weekdays and between 0715 and 1300 on Saturdays"), and where cadets may sunbathe ("only on the roofs of buildings 756A, 735, and the gymnasium sun deck. Sunbathing cadets will insure that they are not visible from the ground during ceremonies").

The penultimate chapter, five, is the longest: twenty-two pages, slightly over a third of the manual. Its title is "Accountability." It specifies where cadets should be during the day, the activities and events of the day they must attend, when inspections are held, how absence cards (which designate a cadet's whereabouts) should be marked, the procedures for reporting on sick call and making appointments with doctors and dentists, accountability procedures in the cadet mess (for example, "Cadets who enter the cadet mess areas late may remain after 'Battalions Rise' to finish their meal. All others will clear the dining facilities without delay"), and the accountability procedures for cadets' participation on West Point's intercollegiate teams.

Chapters three, four, and five together constitute about seventy percent of the regulations manual. The sixth and last chapter, "Authorizations and Privileges," runs a scant nine pages. All cadets are granted certain privileges simply because they belong to the corps of cadets. Other privileges are granted to all cadets in each class who are doing satisfactorily at West Point. Additional privileges which are granted individually are based on exceptional military, athletic, and academic performance. All cadets, for example, can escort women between certain times of the day seven days a week, visit officers on post, hike on post, and take a religious retreat once a year.

A cadet's fourth-class* year is especially heavy on regulations and light on privileges. Plebes are the enlisted men of the corps. The fourth-class system, a modified extension of Beast Barracks, has several functions: to convert and develop a young man into a professional soldier, to identify cadets unable to function under stress, and to provide upperclassmen with leadership training. The system, in fact, imposes an additional set of regulations on freshmen on top of the regular cadet regulations. The regulations weigh heavily enough on all cadets, but the dual set is positively oppressive for plebes. For example, "While using the stairways, fourthclassmen will double-time up the stairs and walk down the stairs, keeping to the side closest to the wall." "When proceeding to or returning from class, fourthclassmen will maintain proper military bearing, not speak unless addressed by a superior, keep head and eyes to the front, and walk at 120 steps per minute." Or, in the mess hall, fourthclassmen will "sit erect, on all of the chair, keeping the head up and eyes directed below the horizontal unless addressed by upperclassmen. If so addressed, they will cease what they are doing and face the upperclassmen speaking to them."† When reporting to an

---

* A freshman is a fourthclassman or "plebe"; a sophomore is a thirdclassman or "yearling"; a junior is a secondclassman or "cow"; a senior is a firstclassman or "firstie." Terms such as freshman, sophomore, etc., are not approved at West Point but are used here for the sake of simplicity.

† On a trial basis for a period of three months beginning March 31, 1974, this regulation was relaxed. Fourthclassmen were allowed to gaze around while sitting in the mess hall, to eat in a more normal fashion, and to talk with each other "in a low tone of voice." Nor were they required to memorize upperclassmen's beverage preferences any longer.

upperclassman's room, freshmen "will remove all overcoats, scarfs, and headgear. The garments will be aligned on the floors against the wall outside the room. The fourthclassman will knock twice, wait for authority to enter, then enter and report, 'Sir, Cadet ———— reports to Cadet ———— for special inspection.' "[1]

It often seems to plebes as if upperclassmen are attempting (in the Academy's words) to "identify those cadets who cannot function under stress" with a vengeance. "If you make it to the latrine and back without getting harassed, you've had a good day," said one unhappy freshman. A typical encounter between a plebe and an upperclassman might go like this:

"What are you doing out here in the hall with your shoes looking like that, meathead?"

"No excuse, sir."

"Buster, you'd better fire back to your room and clean those up, then come around and see me."

"Yes, sir."

One afternoon an upperclassman spotted a plebe whose hair was too long.

"Hey, mister," the upperclassman yelled. The plebe stopped and turned around.

"What company are you from, bean?"

"Sir, I'm from Company F4."

"Do they let you run around with your hair like that in F4?"

"Yes, sir."

"Beanhead, I'm a firstclassman, and I don't have hair half as long as you have. I'll give you exactly thirty minutes to get a haircut and post back to my room. I live right there on the corner, so you know where I am. You've got half an hour, and if you're late you're gonna swing. Post!"

In this atmosphere every fourthclassman looks forward to exam week, not because he enjoys being tested, but because upperclassmen are so busy studying that they have little time to harass plebes.

A fourthclassman's jobs are menial and, even while tinged with humor, are meant to remind plebes of their low status within the corps. At each table in the mess hall, for instance, a fourthclassman acts as the "gunner." The gunner must see to it that his table is supplied with food, silverware, and dishes; he is also responsible for cutting pie or cake and invites constant harassment if he does it incorrectly. Another fourthclassman, responsible for keeping the table supplied with cold beverages, is called a "cold beverage corporal"; a third fourthclassman is the "hot beverage corporal." Both of these men must memorize the upperclassmen's beverage preferences. On a rotating schedule throughout the year, plebes serve as laundry carriers, orderly room orderlies, sink orderlies, trunk room orderlies, mail carriers, minute callers, and linen carriers.

In addition to their duties and their academic work, plebes must also learn vast amounts of "fourth-class knowledge." This "knowledge" is meant to teach men to establish priorities within a short time, to respond effectively under stress, to learn traditions that are part of the Military Academy and army life and to "generate an appropriate sense of curiosity and enthusiasm for matters pertaining to the army, the military profession, and world affairs." In order to stimulate plebes' "curiosity and enthusiasm" West Point packages fourth-class knowledge in nine separate lessons—one for each month of the academic year. The first lesson, largely a review of material learned during Beast Barracks, involves verbatim memorization of Schofield's "Definition of Discipline," army songs and cheers, names of heads of the academic departments, the football schedule (date, opponent, location), the numbers and positions of members of the starting football team, and biographical profiles of the superintendent, the dean, and the commandant. Plebes must also memorize "The Days": movies for the week, weekday sporting events, special programs for the weekend, and the number of days until coming football games, the Navy game, Christmas, five hundredth night, one hundredth night, Mobilization Day, spring leave, and graduation. Plebes also have to memorize the officer-in-charge for the day, the marches for the parades, the reviews for the day, and the menus for all three meals.

Subsequent lessons demand less memorization and have more relevance for cadets' West Point and military careers. For example, plebes must familiarize themselves with quarters regulations, the various uniforms, how to report for formations, and the mechanics of academic work, as well as the requisites of gentlemanly conduct for an officer. But the sheer tediousness of memorization is never completed. Lesson three, for example, requires that cadets learn the order of precedence of U.S. military decorations. Lessons six through nine require cadets to memorize information about the different branches of the army.

Cadets' second year at West Point has a different flavor. The summer for second-year cadets begins auspiciously with two months of summer military training at Camp Buckner on the West Point reservation. "It's at Buckner that we get some responsibility for the first time," said a cadet. As a squad leader for other cadets, a second-year man is responsible for seeing that those in his unit are ready for inspection and parade. He also gives orders while his squad is on maneuvers and is responsible for such tasks as keeping the squad together in proper formation against a simulated enemy.

This taste of responsibility creates problems, however, when a sophomore returns to West Point at the end of August. "It's bad because you don't really have anything important to do," explained a senior. At first, cadets experience a sense of elation about their release from the fourth-class system, but this feeling is short-lived. "You don't have any of the plebe duties, but you realize you don't have any responsibility or privileges, either," said another cadet.

A sophomore's responsibilities are in fact meager, though this might be defended on the grounds that the second year is academically the most difficult of the four. In addition to English, history, psychology, and a foreign language, cadets must study math, physics, and chemistry. In rotation once a month, every second-year man serves as a cadet-in-charge of quarters for a day, which carries with it such responsibilities as sitting in his company's orderly room and answering the telephone, carrying correspondence to his tactical officer from his company commander, picking up company mail, and keeping fire watch (over stone buildings, as one cadet pointed out) while other cadets are at breakfast. Every sophomore also serves as the team leader for two or three plebes during the year. His responsibility is to drill his charges before supper on their memorized material so that they will be prepared when they recite it later for their squad leader.

The symptoms of the second year are outspokenness, apathy, and a certain recklessness. Every sophomore enjoys a special freedom; he has not yet made a solid commitment to West Point, though he has escaped the oppressiveness of plebe year. If he decides to resign prior to the first day he enters classes as a junior, he has no obligation at all to the military. His options still open, a sophomore is more willing to express his feelings. "The yearlings are known as 'the pulse of the corps,'" said one senior. "If you want to know what the corps is thinking, ask a yearling."

But they also seem more at loose ends than other cadets. "There's no sense of direction, no responsibility to anyone but a degraded self," a recent West Point graduate recalled. "Sophomore year is kind of disappointing, kind of a slump year," said a cadet. "Unless you get involved in other stuff, it's a depressing year." "Other stuff" are activities such as intercollegiate sports, intramural sports, or interest clubs—the cadet glee club, photography club, and so forth—which enable cadets to escape West Point. For many sophomores, however, "other things" means challenging or breaking the regulations. "Second year was socially regressive, but we had to express ourselves somehow," said a senior. "I've never done so much messing around in the halls, had so many water fights, or drunk so much on the roof." Rule-breaking, in fact, is the hallmark of the second year at West Point. Drinking on the roof of the barracks, sneaking out after taps, missing required formation, and not keeping one's uniform in order are attempts to express autonomy and break through a feeling of uselessness. A second-year cadet is not risking so much as other cadets because he does not have so much to lose. One sophomore said, "The yearling year is supposed to be contemplative, a year to decide whether you really want an army career and what you want to do if you stay in. But that gets lost somehow."

While a cadet's sophomore year is not rich with privilege, it represents a vast improvement over freshman year. Sophomores are "authorized" to take trips off-post, to fish and hunt, to eat supper with staff and family members on Friday evenings, and to take spring leave— seven days—for the first time. They may also visit each other in their rooms during evening study periods and take one weekend of leave each term.

Third year at the Academy also begins with summer training: AOT or Advanced Orientation Training. On the basis of their grades and preferences, cadets receive assignments to major army posts throughout the United States and Europe: Fort Knox, Fort Benning, Fort Sill, Fort Bliss, Fort Belvoir, West Germany, and Hawaii. "You have a real command position for the first time," recalled a 1972 graduate. "We were acting second lieutenants in charge of a whole platoon for two months," a cadet recalled. "I served as a forward observer for an artillery unit in Alaska. I was finally able to escape the West Point label and feel like an officer. I had a real job to do. When I left, I felt I'd earned the respect of the officers and their men."

The minute a third-year cadet steps into the classroom at the beginning of the academic year, his fate is sealed for the next seven; that act commits him to two more years at the Academy and five more as a commissioned officer in the army. If a man decides to resign between his first class of the junior year and graduation, he is obligated to serve two or three years in the army as an enlisted man. "There's a saying at West Point," said a junior, "that walking into class the first day of third year is like breaking a mirror. Both bring seven years of bad luck."

But third year also marks the point at which cadets begin to assume responsibility in the cadet chain of command. "Junior year's pretty good," commented a high-ranking senior in 1972. "You're only one year away from graduation, and if you get selected as assistant executive officer or assistant first sergeant in your company, you know you're in line to get something pretty big in your senior year." Every third-year cadet also serves as squad leader for at least three months. In this role, with ten men under his command, his duties consist of preparing fourth-classmen for inspection and parades and seeing that plebes are in bed on time. "But the problem," commented a rather disillusioned senior in 1972, "is that you're required to reinforce the very system you've learned to despise. And, really, there's no responsibility as squad leader. You're merely involved in passing down orders from above."

Everyone agrees, however, that the junior year at least means more "authorizations"—as privileges are called at West Point. Every third-year cadet may take three weekends of leave a term, attend the week-night movies after first term if his academic work is in order, and drive another person's car on post with the owner's permission. Also, cadets participate in two important Academy ceremonies for the first time in their third year: five hundredth night and the ring ceremony. Five hundredth night—five hundred days until graduation—is celebrated by a banquet and dance. Juniors are allowed to escort their dates on Friday evening and Saturday morning and to hold open house in barracks. "It's all a morale booster," commented a senior. The ring ceremony, which takes place the end of the third year, marks the cadet's full entrance into the corps. "It's kind of like joining the alumni," commented a recent graduate.

After the seniors graduate in June, half of the new senior class leaves for their "first-class trip." The other half stays at West Point in order to train the new cadets who enter each July. On the first of August, those who went on the first-class trip in July return to West Point to take over the training of the cadets, while those who remained then leave on their trip. The first-class trip, a brief visit to each of the headquarters of the combat arms branches of the army—Fort Sill for field artillery, Fort Knox for armor, Fort Benning for infantry, and so on—is meant to familiarize cadets with the branches in preparation for their branch drawing in February. Branch selection is, in fact, one of the big events of senior year. The corps of engineers is always the first to fill its quota, followed by military intelligence. As a general rule, the higher ranking cadets in the class select the corps of engineers, still the elite branch to West Pointers. Infantry, the largest army branch, also obtains the largest number of cadets—approximately a third of each year's class. Cadets make their choice in rank order on the basis of their order of merit within their class, which means that the first man gets his choice of branch, while the lower ranking men take whatever branches are not yet filled.

Seniors assume the most responsible positions in the cadet chain of command as regimental or battalion officers, company commanders, or company executive officers. Company commanders often complain that their job, on top of academic work, is too strenuous, especially if their tactical officer is insecure and meddlesome. One cadet who served as a company executive officer his senior year said, "We had a good tac. He told us, 'I'm going to leave you guys alone unless I get reports from the officer-of-the-day or other tacs. As long as you're doing okay in company drill and marching, you can do it yourself.' That's the way it worked. We got the company the way we wanted it. We watched out for each other and policed ourselves."

Another man, however, who served as a company commander for two terms his final year had a less happy experience. "I was D [academically deficient] for several weeks during first-class year because I was so busy trying to appease my tac, trying to keep him off the company's back. I think it was a worthless year for me. I had responsibility but no authority. The tac would come by on inspection every morning, usually without warning us first. The closets [in the cadet rooms] had to be open and he would find all kinds of things wrong, like shoes not being shined properly, the room not swept well enough, civvies in the closet, cellophane on white shirts that had been returned from the cleaners, a trash can with something in it, books not arranged in order—the tallest books have to be on the left side of the shelf and descend in size so that the smallest books are on the right—the broom not under the left-hand bed with the bristles toward the ceiling. If things were bad enough, he'd get an angry look on his face, would say, 'The company isn't prepared for inspection,' and go back to his office and sulk or something. Sometimes he'd slug me with an eight and eight."*

Their last year is probably the happiest of the four at West Point for seniors not assigned to command positions. "If you don't have responsibility, you have lots of time and most weekends free. It's a pretty good life," said a cadet. In the course of senior year, many of the privileges which a cadet surrendered on the day he entered West Point are returned to him. Seniors are granted more free weekends each term—six for everyone, and more if the cadet makes the dean's, commandant's, or superintendent's list. He also has the opportunity to dine in the

---

* Punish the cadet by "awarding" him eight demerits and eight hours of confinement.

officers' club after first quarter; the privilege of drinking with staff, faculty, coaches, and of wearing civilian clothes of officers' homes; the authority to leave post until taps on weekends and holidays; and—perhaps most important to cadets—the authorization to own a car and keep it on post three months before graduation. This means, as one officer observed, "A cadet has to put four years of living into three months." But a car, said a 1972 graduate, "is a symbol of the fact you'll be out soon, a symbol you've finally arrived. It also means you can get out of the place any time you want." Cadets, with some justification, complain, "At West Point they take away your rights and give them back as privileges."

At the end of his four years at West Point, every cadet is thoroughly indoctrinated into the military way of life to the extent that the civilian world has assumed another reality altogether. This estrangement begins immediately in Beast Barracks and is well under way by the end of a cadet's freshman year. "Plebe year is strange," recalled a senior. "It turns you into a social eunuch. You're separated from everything you've known before. In your isolation, the only force that is applied to you is the force that the Academy chooses to apply. After a year here, West Point becomes your home. You feel safe and secure here and insecure on the outside." Indeed, the fact that so many competent high school students feel "insecure" on the outside after a year at the Academy is testimony to the effectiveness of West Point training. The reasons are evident: every cadet is treated harshly; no allowance is made or recognition given for his past achievements; he is isolated from his previous life and sources of support—family and friends. A cadet's self-esteem, based previously on his achievements and reinforced by the approval and support of his family, friends, and teachers in the civilian world, is crushed by the Academy. Only by adhering to the military way can he regain it. In the authoritarian West Point world, where mistakes and punishment are almost synonymous, cadets find safety, approval, and well-being by following the regulations and by doing as they are told. Cadets are no more immune to conditioning than the rest of us. Once they adapt to military life, it is no wonder the civilian world, with fewer regulations, less order, and more opportunity of choice, becomes a territory in which they feel out of place and even apprehensive. In 1972 a senior commented, "Every cadet is twenty-eight years old. Because West Point is so different from college, we don't feel comfortable around people our own age anymore. Yet we're not adults either. We're somewhere in between." Two years later, in 1974, another senior observed, "I'm so isolated from the outside world that I feel uneasy whenever I go

39

out. West Point has been good for me because I'm good at functioning by habit; I can't do that when I go into the civilian world. But I'm not sure it's good if you can excel at West Point and still not have your stuff together when you go out, if you know what I mean."

This cadet's experience is common and inevitable. As cadets accommodate themselves and as their ties to the civilian world grow progressively weaker during their four years at West Point, the Academy takes on the dimensions of a reality as substantial and important as their civilian world ever was.

5

CADET LIFE

Each year at West Point has its own emphasis. Cadets undergo indoctrination and socialization into the military world during their first year. Their sophomore year is a period of restlessness, rebellion, and strenuous academic work. The third year is preparation for leadership, while the fourth brings leadership responsibilities in the cadet chain of command or, conversely, more freedom than a cadet has enjoyed before at West Point. This is only the skeleton of the West Point experience, however. The flesh is the quality of day-to-day life, the atmosphere in which cadets must function at West Point.

Regularity, Order, Discipline could be the Academy's motto as well as Duty, Honor, Country. Only some future sensitive cadet diarist will be able to describe Academy life as the intensely felt experience it is, if he can find the time. The first reality of Academy life for every cadet is that his time is heavily scheduled. Up by 6:15 A.M. and in ranks ten minutes later, he marches to breakfast at 6:30 and returns to barracks about twenty-five minutes later. Classes begin at 7:45 A.M. Lunch begins at 12:10 P.M. Afternoon classes start at 1:05 and end at 3:15; parades or intramural athletics follow a short time later. Supper begins at 6:15, the evening study period an hour or so later. Taps come at 11:00 P.M. for everyone and lights must be out by 1:00 A.M.

All barracks rooms must look alike, and each cadet follows a "Barracks Arrangement Guide," complete with photographs, which shows where blankets should go, how clothing should be arranged, and how belts should be placed in drawers. If windows in the room are open, they must be open to exactly the same height. All cadets are authorized to have two knickknacks (small objects such as statues or figures), one pencil holder, a calendar, and one picture on top of their desks. In the fall of 1972, the new commandant, Philip Feir (when he first arrived at the Academy cadets were saying, "We have nothing to fear but Feir himself"), was reportedly bothered because cadet roommates had different colored toothbrushes; it distracted from the "uniformity of the corps." Extra books can also detract from uniformity. One sophomore who took an interest in English and bought several unrequired books on nineteenth-century literature was told to get rid of them by his tactical officer. When asked why, the officer said, "Nobody keeps them and they don't look military. Besides, what would happen if everyone had extra books?"

Cadets are watched very closely. Asked one, "If cadets are such mature and responsible individuals, why are we *always* given something to do? Everyone wants to keep his eye on cadets." Although cadets are presumably on their honor to tell the truth at all times, tactical officers have been known to ask a cadet where he has been, then check up on his story immediately afterward. And some officers, for reasons of their own, seem to make a fetish out of catching cadets for various infractions of the regulations. One lieutenant colonel, cadets swore, used to drive back and forth in front of Grant Hall in order to catch and punish those who failed to salute his car. Cadets also felt the same officer came to movies in order to catch them creating a disturbance or holding hands with their dates.

Almost every officer naturally feels it part of his duty to enforce the regulations. "You've got to. If there are rules, they must be upheld," said a West Point major. This attitude meant no rest for cadets. One morning while taking sick call, I saw a cadet who had a severe headache. After we talked a few minutes, he told me he was very unhappy at the Academy. His headaches had started four days earlier, on a Saturday, when he was walking with a girl up a steep hill near the cadet chapel. Because she was tired and the slope was steep, she took his arm. A station wagon with a man and his wife and three children cruised by and stopped. A major in civilian clothes got out of his car and yelled, "Come over here, young fellow." He then dressed the cadet down, in front of his date, for a "public display of affection." "It ruined my weekend," the cadet said.

The extent to which cadets are sometimes watched is shown by an example provided by a West Point doctor. One evening a sophomore cadet was brought to the emergency room. The doctor on call asked him what the matter was. The cadet answered, "Cadet Jones, sir, company A2." The doctor asked again. The cadet gave his name and company for a second time. The doctor, baffled, said, "You're not a POW. I'm a doctor and you can tell me what the problem is." The cadet gave the same answer a third time. At that point, the physician decided to call the cadet's tactical officer, a major, and ask whether anything had happened in the cadet's life lately that might throw light on this strange behavior. The tactical officer said, "I don't think so, but I'll look in his file." The officer returned and said, "There's nothing in his file except a couple of reports that he picks his nose in class." It turned out that the cadet had suffered a brain concussion in intramural football, from which he eventually recovered.

In this tight, controlled environment cadets are also constantly evaluated. Every week the nine squad leaders in each company must prepare an evaluation on all the members of their squad. These reports, written up in the squad book, are submitted to their company commanders and to their tactical officers. Reports are concerned with how men are getting along at the Academy. A typical report might read: "Mr. Smith doesn't seem to be adjusting to the Academy too well; he doesn't get along well with other cadets, and he is deficient in three classes." Or, "He seemed to have his room in order for Saturday morning inspection and looked sharp in ranks." Every semester cadets are asked to rank each other in numerical order on military aptitude, a man's ability to lead a unit in accomplishing a mission while maintaining morale and discipline within the group. Those who end up at the bottom in the first ratings have an enormously hard time raising themselves in subsequent ratings. Plebes are graded every day in most of their classes. Cadets are graded less frequently as they approach graduation, but even seniors are still graded at least twice a week in most courses. Cadets always know exactly where they stand each week in every class they take; grade averages are posted publicly.

The oppressiveness of West Point is hard for an outsider to comprehend. The smallest and most trivial details of life receive daily attention and correction. Cadets, for example, must salute officers' cars, identified by blue bumper stickers, even if the officer is wearing civilian clothes; not to do so is a punishable offense. During inspection, it is not unusual for an officer—usually a major and sometimes a lieutenant colonel—to look in a cadet's soap dish, check carefully under his bed, or empty out his laundry bag (a favorite hiding place for liquor). "Imagine a thirty-six-year-old man doing that!" exclaimed one cadet. But a major in the mathematics department attempted to place the matter in perspective. "Cadets are amazed because many of them don't understand that military training here demands the highest standards. When an officer says, 'I want that soap dish clean,' he means it. One rarely sees these high standards maintained in the army, but, by God, an officer who has gone to West Point, even if he doesn't enforce them, will never forget what the standards are. That's what this place is all about."

"Standards have their place," said a sophomore, "but around here they're carried to lengths which are ridiculous. There is a saying that West Point isn't the army. That's one of our salvations. The army couldn't be as petty as West Point. Last week the assistant commandant came around on inspection. My roommate had a sheet of acetate on the top of his desk with a *Playboy* picture underneath it. We're not supposed to have acetate on our desks; it's been a big deal around here for two years—supposed to be bad for the eyes or something. But when the colonel saw the acetate and the picture, he went berserk. He tore up the picture and threw the acetate on the floor. You have to ask yourself: Is that what I'll be like if I stay in the army? Is that the kind of person I'll have to associate with? Is that what the army does to people?"

A senior said, "General Feir really stresses table manners. He comes to the mess hall himself to make corrections. He stopped one of the guys in my class a while ago and told him that the proper way to put peanut butter on bread is to lay the bread on the plate first, then spread peanut butter on it."

Punishment is a more overt way of regulating cadets' behavior. "Quill"—cadet slang for demerit slips—is handed out more freely than anything else at West Point. All cadets know the consequences of misbehavior, which may have been nothing more than sticking one's head out the window or holding hands with a girl in public. Hand holding, which is considered poor military behavior, warrants seven demerits if it occurs indoors, eight if it occurs in a public place. A "public display of affection" which takes the form of walking with a hand around a young woman's waist brings ten demerits and fourteen punishments (a punishment consists of marching for one hour with a rifle in a specific area near the barracks). Eight demerits and eight "punishments" are "awarded" to a cadet who cannot control a formation he is leading. Intentionally missing a class will invite twenty demerits, forty-four punishment tours, and two months' confinement to one's room. Upperclassmen are allowed 0.6 demerit per day, which means thirteen demerits a month. If a man receives fifteen demerits in a month, for example, he has to work off the two extra demerits by marching in the area for two hours; or, if he is a senior, by staying confined to his room for two periods (approximately three hours of confinement composes one "period") during the evening or on weekends. The system is more lenient with plebes, who may accumulate twenty demerits a month each before they are forced to march.

Though admittedly an exception, one officer passed out one hundred "pieces of quill" in one weekend alone in 1972. At a regular time each week, "punishments" are awarded to offending cadets. A typical "slug sheet" reads like this:

Awarding of punishments:

1. Cook, James K. 27332 (210-52-9136), CDT PVT Co A, 1st Regt., Class 1973

   Type Board: Cadet Regimental Board

   Offense Code: 360

   Offense: Unauthorized possession of material not within USCC ethical standards, i.e., two "dirty books"

   Punishment: Ten (10) demerits and fourteen (14) punishments

   Release Date: NA

2. Knowles, Richard C. 25210 (048-52-5412) CDT PVT Co B, 4th Regt., Class 1974

   Type Board: Cadet Regimental Board

   Offense Code: 751

   Offense: Displaying belligerent attitude toward superiors, i.e., making obscene gesture to firstclassman through window

   Punishment: Fifteen (15) demerits and twenty (20) punishments

   Release Date: NA

A close relation to the constant supervision and discipline of cadets is the Academy's unwillingness to allow them any real responsibility. Donald Cantley, a West Point senior who subsequently resigned before his graduation in 1972, wrote in *The Pointer,* the cadet magazine:

The cadet lacks determination power over most areas of his life: the questions of when, where, how and even whether he eats; when, where, and how he sleeps—to stick to the more basic functions—are all determined *for* the cadet. The smothering security that this sort of "care" provides is in large part responsible . . . for cadet immaturity: treated like an irresponsible child, he quickly learns—remembers—how to behave like one.

45

The central, ironic paradox of Academy life is that the institution attempts to "build" leaders by denying them room for individual choice, thought, and initiative. Cadets are aware, bitterly so, of the paradox; they resent that—ostensibly for socialization purposes—they are expected to take seriously such responsibilities as insuring exact shoe alignment but are effectively denied any real leadership—decisional—role.[1]

Or, as one cadet more succinctly put it, "It's hard to be a man here."

The manner in which West Point handles cadets' finances is but one example of how responsibility is sometimes taught at the Academy. A recently graduated officer, about to start his career at Fort Benning, said, "We get 256 dollars a month,* and it's a good thing they don't let us have it all. You have a checking account and a fixed account. The treasurer handles the fixed account, and out of that automatically comes the money for uniforms, books, candy, the magazines and newspapers we are required to take in the company: *Newsweek, Atlantic, The Wall Street Journal, The New York Times*—there has to be one *New York Times* for every three cadets. One hundred ten to 120 dollars is sent to our checking account. If you overdraw, you get slugged eight and eight. But usually the bank tries to cover you. The Academy invests our money; I don't know how or in what. All I know is I get 1,700 dollars in a lump sum at graduation and a 4,000-dollar loan from the bank to pay for my car."

The stories of cadets' dependence are amusing but also faintly pitiful. A group of cadets anticipating a trip to Philadelphia did not plan to take money because they assumed the officer-in-charge would pay for everything. When his car broke down the week before graduation, a senior called one of his instructors and asked him what to do about it. The day after graduation, a West Pointer called his tactical officer for advice when he found himself locked out of his parents' motel room.

Cadets frequently complain that they are accountable but not responsible. What this means is that they are not trusted to make decisions on their own. In the fall of 1973, the cadet brigade commander, using his own judgment in the face of new regulations which explicitly urged cadets to use "common sense" in making decisions, told the corps that it was permissible for cadets to wear Corfam shoes and wire-rim aviator glasses (black Corfam shoes, which require minimal upkeep, and wire-rim glasses, which presumably distract from the uniformity of the Corps, have always been proscribed at the Academy). The next day, however,

---

* Cadets now earn $305.45 a month.

the commandant revoked the brigade commander's decision. A short time later, the same brigade commander decided that firstclassmen could miss Friday supper if they happened to be escorting a woman then at that time. Again, the commandant said "No." "When the brigade commander, who's a good guy this year, started making decisions on his own, the commandant came in and cut him off," one cadet said. "Now he's afraid to make any decisions. What little independence he had is gone." In a recent interview with a *New York Times* reporter, Brigadier General Hoyt S. Vandenberg, Jr., the commandant at the Air Force Academy and a 1951 West Point graduate, clearly defined the military concept of responsibility. "Policy is made by high authority. What we mean by giving them [cadets] more responsibility is that we hand down policy and they carry it out."[2]

The most common reactions to being "accountable but not responsible" are apathy and cynicism. Unable to affect the system, many cadets simply succumb to it. "All you do is lock yourself into the schedule here," said one cadet, "and you've got it made. You don't have to think about anything or worry about meals, clothes, or entertainment. It's all provided. You just do what's on the schedule."

Apathy is also expressed in that last refuge of the unhappy—sleeping. In their crowded schedule, it is hard to know when cadets find time to sleep, but they do find it: between breakfast and their first class, between athletics (or a parade) and dinner, and of course during weekends. "An hour of sleep is an hour away from West Point," said a sophomore. A senior observed, "Sleep is the main recreation here; it's a symbol of escape. It means not thinking and is a way to remove yourself easily and quickly from the surrounding environment."

Apathy shows, too, in a routine, spiritless obedience to the schedule. "If you just do what you're told," said one cadet, "you can get things out of the way faster and have some time for yourself. It's just too much of a hassle not to do something you're told to do." Another cadet admitted, "If we didn't have parades on Tuesday and Thursday, I wouldn't know what day it is half the time." For at least some cadets, apathy reaches its height during fall term of senior year. "This is the worst time," exclaimed a senior. "Eighty percent of the guys feel like I do. We've been here almost four years, seen what it's like, and what do we have to look forward to? Five more years in the army."

Cadets turn apathetic not only because they feel powerless, but also because they feel overwhelmed. "It's hard not to get apathetic here," said a cadet who graduated near the top of his 1972 class. "There are just too many things to do. For most of us it's just easier to sit back and let it

47

flow over you. Being at West Point is like being in a room filled with eleven balloons; you're responsible for keeping them all in the air at the same time. By the time you get around to the eleventh, the first is almost on the floor, so you have to tend to it while the others continue their fall. It gets frustrating because you can never concentrate on any one thing for any length of time."

The cynicism and disaffection that claim some cadets after four years of powerlessness, frustration, and anger were clear from the remarks of a graduating senior in 1972: "The only reality at West Point is the myth. I don't complain much to outsiders, because I'll look like a fool for putting up with it. Why should I destroy the only thing I've liked about West Point?" A 1962 graduate who had earned a master's degree at Princeton and returned to teach at the Academy agreed: "I got my education when I went back to Princeton. When I was here, I just wanted to get through. The symbol—what West Point stands for in the minds of other people— is far more important than the reality."

It should be said that not *all* cadets are apathetic and cynical. Some seem to identify completely with West Point and have very few complaints about it. "I think lots of them are guys who were in the army before they came to the Academy," said a cadet. "Some of them wanted to come to West Point before and couldn't get in; they appreciate it after being an enlisted man." Also, cadets who are most successful at competing seem to enjoy West Point. "The pattern is set during plebe year," said a cadet. "If a cadet sees he can do well, then he's set. If he doesn't, he turns sour." Most cadets are very competitive when they enter West Point. Further success in academic or athletic competition usually sends cadets in an ascending spiral upward to higher leadership positions. On the other hand, as one cadet said, "Competition alienates a lot of people. It just beats them on the head, especially in academics, and they decide not to strive at all. They tend to form cliques and are critical of everything that West Point and the Army does."

It is very difficult, however, to avoid competing at West Point. Competitiveness is a fact of daily life at the Academy. For example, privileges largely depend on a man's ability to compete successfully. Any cadet who makes the dean's list (the top thirty percent of the class academically) receives extra weekend leaves and leave time for extracurricular trips. The commandant's and superintendent's lists, both instituted in 1973, also carry extra privilege. Any senior who makes the commandant's list—recognition for "military excellence" based on a cadet's aptitude ratings, his physical education grades, his grade in military instruction, and the number of demerits he has received (the fewer the better)— is allowed to return from weekend leave at 10:30 P.M. instead of 6:30 P.M. on Sunday and may miss dinner and go beyond post limits on Friday until 10:00 P.M. A junior who makes the list is authorized to miss dinner on Friday and escort a woman until taps that night, while a sophomore may attend athletic events on weekdays. The diligent senior who achieves the superintendent's list (makes the dean's list and the commandant's list simultaneously) is authorized unlimited regular weekend leaves and five long weekend leaves each term. Juniors on the list are granted five regular and three long weekend leaves a term, while sophomores may take three and two respectively.

Probably nowhere is competition stressed more than in the realm of athletics. Within the corps of cadets, the basic units of competition are the companies, which vie with each other perpetually, not only in marching, inspections, and academic work, but also in intramural sports. Cadets who belong to companies which attain superiority, especially in athletics, receive special privileges. "Last year," said one cadet, "when one of the companies won the regimental championship in intramurals, the tac decided to reward them. He let them arrange their rooms the way they wanted and let them put posters on the backs of their doors. Seniors in another company who won the regimental championship last year only had to have their rooms in order for PMI all year. To pass PMI— afternoon inspection—your room just has to be in a semblance of order, not like it has to be for AMI or SMI [morning inspection or Saturday morning inspection, respectively]. Not to have AMI or SMI is a big thing."

The competition at West Point is tough enough so that any cadet who is physically inept receives more harassment than one who shows athletic ability. "If you don't measure up physically, you get poor aptitude ratings," said one man, who then added with unassailable logic, "You lose games if you've got a poor athlete on your squad. There are enormous pressures to compete—from the athletic department, from your peers, from your tacs. Five out of thirty yearlings in one company were thrown out two years ago by the physical education department. It's for sure that athletic ability has to do with rank in the corps, too. Boxing is really stressed here, for example. A good friend of mine who almost won the brigade championship in boxing is a battalion commander this year. One of the regimental commanders, two other company commanders in my regiment, and a regimental commander last year also boxed."

Two social scientists at Dartmouth, John Masland and Laurence Radway, have pointed out that the service academies invest more time and money in athletics because sports develop special qualities regarded as essential in a fighting man: an aggressive, competitive spirit and the strength and endurance to overcome severe physical hardships.[3] Competitive athletics "on the fields of friendly strife" are also a more or less civilized substitute for war.

The Academy also knows that sport is a form of sublimation, a way of relieving stress and channeling feelings in approved military ways. "When I'm around the company for any length of time and going to classes every day," said a burly 200-pound junior, "I can't wait to get to intramural football and smear someone."

West Point invests heavily in sports for other reasons as well. The Academy expects its athletics program to enhance "the reputation and traditions of the military Academy by the quality of performance and conduct in practice and in competition, at home and away." In a physical world like West Point where, as one officer said, "the elemental truth of success is looking good," the performance of Academy teams becomes a reflection of the institution itself. A winning football team is a justification of the Academy and its system. A losing team, however, is a disaster that plunges the Academy, especially the officer corps, into gloom and bitter self-searching.

But the gloom that attends a losing football season has deeper roots. A major in the English department expressed them: "Cadets and officers give up a lot at West Point; they sacrifice. Unconsciously, they want something for what they've given up. Because they've sacrificed so much, they think that God is, or should be, on their side. A losing football team means that they're not being rewarded, and they feel bitter about it. They say, 'We, even as underdogs, should win because of the way we've lived.' Some officers I know are incurably optimistic about West Point's football team. They place bets with friends on the most hopeless games. They can't accept the idea of losing." The fortunes of the football team are so closely linked with West Point's image that, as one cadet observed in the fall of 1970 (a notoriously bad year for the team), "If anything changes West Point one way or another, it'll be two losing football seasons."

West Point finally fired its football coach, Tom Cahill, in December 1973. The football team had staggered through the season, losing nine straight games. Cahill's fate was inescapable, however, when Army dropped its tenth and final game on December 2 to archfoe Navy, 51-0, the worst defeat ever suffered by either team in their long rivalry. One can only imagine the collective bitterness expressed at West Point that night. As Coach Cahill received news of his dismissal from the superintendent, he asked, "Is this because the image of the football team hurts the image of the army?" General Knowlton said yes. Cahill replied, "The problem with the football team is that it is suffering from the image of the army."[4]

The amount of athletic activity at West Point is extraordinary. On May 6, 1971, for example, cadets participated in eleven different athletic events with nine other colleges at five separate locations. One of the reasons for the great popularity of sports is that it allows young men an escape from regimentation and conformity. (All extracurricular activities at West Point serve this function. A major attraction of any special interest group, especially the glee club, is the opportunity it affords cadets to get away from the Academy.) "Everyone loves the athletic parts of West Point," commented a senior. "Maybe it's like taking dope to escape something you don't like."

Athletics is one safety valve for cadets at West Point; rule-breaking is another. Rule-breaking may seem a curious method of coping with the Academy, but it is a method. Any number of motivations lie behind rule-breaking. Many cadets feel that the regulations are trivial and unrealistically restrictive; some break rules as a form of risk-taking, while others break them out of anger. "I've broken the regs when I've been pissed off about something just to get back in some small way," said a junior.

More often than not, regulations are stretched to the limit or broken in exceedingly subtle ways. Cadets will let their hair grow right to the three-inch limit, or wear disreputable clothes, especially articles whose disrepair cannot be easily detected—shoes with holes in the bottom and shirts frayed beyond repair (yet hidden under coats) are common articles of expression. It is very popular to get out of a required formation in any way possible, and the ability to miss a parade is given high marks by peers. Putting something over on the company tactical officer is also popular. If a regulation, say, drinking in barracks, has been broken and everyone in the company except him knows about it, cadets feel they have scored a point. More overt forms of cadet rule-breaking, notable for their boldness, include sneaking out after taps and smoking marijuana or drinking in their rooms.

Drinking and the use of drugs are of course violations of the regulations at West Point. "Everyone drinks," said one cadet. "It's just a matter of who gets caught." As in the civilian world, the penalty for drug use is much more severe than for drinking. Drinking, often orgiastic among cadets on weekends, results in rather drastic punishment if it occurs on post, but the use of drugs invariably results in expulsion. It is impossible to know how many cadets have experimented with drugs at West Point.

Some cadets place the figure at ten percent, others at forty percent. The use of marijuana is common, and a number of cadets have used hallucinogenic drugs, particularly mescaline. Narcotics and barbiturates are almost unheard of at the Academy, however. Never has a cadet been seen in the hospital emergency room at the Academy for a bad drug experience. This does not mean that bad trips never occur, that drug use is not high among certain small groups—a drug-of-the-month club existed among several cadets in one company in 1971—but, rather, that cadets do not come for treatment because they fear punishment from Academy officials.

Cadets also deploy other strategies to preserve their psychological equilibrium. Humor is crucial for survival. "You've got to realize it's the system, that everyone gets treated the same," said a junior. "The minute you start taking it seriously, you're dead." Humor serves to put distance between oneself and the situation at hand; it serves as a buffer between a cadet's self-esteem and the environment.

A large part of the cadets' monthly magazine is devoted to humor. One article, which appeared in the December 1971 edition of *The Pointer,* was called "The Uncle Greentrou Show." "Greentrou" obviously referred to the green trousers of the regular army officers. The article was a satire on the officer-in-charge of the day, who was depicted as a silly individual who spoke baby talk and who took obvious pleasure in writing up cadets for violations of the rules. The first photograph showed a cadet dressed up as an army captain, carrying a large delinquency report pad under his arm. The caption over the picture read, "In an answer to TV's Sesame Street and the CBS Report, *The Pointer* offers its own journey through Fantasy Land. . . . Close your eyes, think pure thoughts, and come with us to the world of gray land." Then Greentrou speaks: "Hi, boys and girls. I'm Uncle Greentrou, your friendly man in Grayland. Follow me and my magic paper as I see what mischief is being made in Grayland."

The article was a serious commentary on the way cadets feel infantilized by officers at West Point and on the fantastic West Point world itself, where appearances are more important than substance. For example, in the article the room in which the officer gleefully writes up the delinquency reports is called the "Happy Base Camp." The last picture in the series showed Greentrou surrounded by small children, with this caption underneath: "Here I am with all my friends. Maybe someday they can come to West Point to live and have a happy four-year existence. Think of all they will learn! Don't worry about my show going off the air. . . . It's been running since 1802."

A cartoon in an earlier edition of the same magazine showed a large screw atop the Civil War monument at Trophy Point instead of the usual Angel of Victory. The officer-in-charge of the magazine, a major, was "relieved" of his position because of it.

Friendships are as important as humor for survival at West Point. A cadet without a friend is a cadet in serious trouble and one who inevitably resigns from the Academy. Pressures at West Point are too intense for a man to go it alone very long; everyone needs the company, support, and help of others. "You need someone to laugh and joke and make good conversation with," said a senior. "Then things don't seem so bad."

Although many cadets are convinced that their West Point friendships are the most intense they have ever known, I wondered at times how deep their relationships with each other really went. Their mutual experiences of deprivation and adversity no doubt encouraged deep loyalties, but it seemed to me that the military discouragement of individuality and self-expression provided thin ground for the growth of personal relationships of any depth. Training at West Point is schooling in self-protection, in saving face, rather than an education in self-revelation and the possibilities of friendship. About personal relationships, one 1963 graduate had said, "You develop a way of dealing with others by not exposing yourself at all. I think cadets seldom get close to each other. You learn to interact—you have to if you want to survive —but it's on a superficial level."

Hazing, defined by the Academy as "any unauthorized assumption of authority by one cadet over another cadet whereby the latter should suffer or be exposed to any cruelty, indignation, humiliation, hardship, or the oppression, deprivation, or abridgment of his legal rights," is now officially prohibited at West Point, though it was once a favorite method of coping with boredom and discontent among upperclassmen. Gone are the days when a plebe had to run up and down stairs while wearing a rubber poncho, hold a rifle out in front of him until it dropped, perform pushups or situps in the hallways, or "brace." An article written by several doctors at the West Point hospital in 1970 revealed that "bracing" —forcing cadets to stand in an exaggerated position of attention with their shoulders thrown back and their arms pushed down toward the ground—caused nerve damage and arm paralysis in 138 cadets over a six-year period.[5] Although the paralysis was temporary in all cases but one and cleared up within six weeks, it left cadets temporarily unable to raise their arms more than forty degrees from their body.

Although hazing is forbidden, harassment is not, as all plebes know. Upperclassmen who wish to discharge and displace their frustration or anger onto plebes do so by ordering them to come around for special inspections, by giving them extra duties, or by having them recite frequently at the mess tables. And, as one senior put it, "If we beat Navy at football, plebes have a much easier time of it between Thanksgiving and Christmas than if we lose. Plebes carry the brunt of upperclassmen's frustrations, too, if the football team is having a losing season." Harassment can become very coarse during Beast Barracks. Some upperclassmen once held "manhood sessions," in which new cadets were asked whether or not they were virgins. At mealtimes, new cadets have been asked to describe the "grossest thing" they have done with a woman. If the report is not satisfactory, the failed cadet is deprived of his dessert and ordered to invent scenes from his imagination for the pleasure of the table.

I'm aware that my description of Academy life makes it sound grim and joyless. For the most part it is. Cadets, naturally the harshest critics of West Point, complain most frequently and bitterly about the regulations. "They prohibit everything," said one. But this does not mean that most cadets walk around looking depressed or despondent. On the contrary, cadets are vigorous, healthy young men whose naturally high spirits cannot be subdued even by the Academy; also, a lasting depression in men this age would be unusual and abnormal.

But the rules and regulations, inspections and evaluations, frustrations and pressures, and the unending stress on competition inevitably take their toll. The disaffection that afflicts many cadets and its expression in apathy, cynicism, rule-breaking, and harassment are the price West Point must pay to perpetuate its system of turning boys into soldiers.

6

CADETS WHO STAY AND CADETS WHO RESIGN

If the Academy is as oppressive as I have portrayed it, why do cadets stay? For a variety of reasons which form a web so intricate that anyone who seeks to isolate a single strand is soon discouraged. The reasons which bring a cadet to the Academy in the first place—prestige, status, the promise of a career, a free education—also serve to hold him there. A cadet remains, also, for reasons that only become apparent once he has been at the Academy awhile: strong friendships with other cadets, athletics, the fear of being thought a "quitter" if he resigns, a feeling that he has an investment in the Academy ("When you've been through so much and put so much into the place, you hate to throw it away," said a junior), and pressure from parents to stay. Others depend on the structure that West Point provides. As one cadet put it, "I really don't know how I could have made it on the outside the way I have here. If I didn't have someone on my back all the time, I wouldn't have done anything."

In addition to these considerations, every cadet has his own personal reasons for remaining at the Academy. A senior commented, "I really enjoy athletics and some of the classes. I have a good English professor and a really good thermo [thermodynamics] and fluids [fluid mechanics] professor. Once you get into advanced classes, you get treated differently.

The officers are more like teachers than officers; they're interested in you and whether or not you understand things. Besides, it's amazing what a person can adapt to. Unless you get caught, the regs don't interfere with living; you learn to ignore small annoyances. It's true that they take away all your rights and give them back slowly as privileges, but you appreciate them more when you get them."

A recent graduate who considered resigning several times during his Academy career came to dislike West Point intensely by the time he had graduated. But he did not resign, because "I wanted to prove I could do it. Living on the West Coast, you get a distorted view of West Point—like it's a place to be treasured. I wanted to prove I could get something neat like that. It's the idea of fulfilling a childhood dream. I'd wanted to go to West Point since I was six years old. I know it sounds incredible, but I first got the idea after watching *The West Point Story* on TV."

"Another reason I stayed," he went on, "was that my parents and their friends still think of West Point as one of *the* places to go. I didn't want to disappoint people by quitting. I couldn't face the idea of going back and facing them if I left. It's especially hard when you're from a small town."

This cadet agreed that life was rather grim at the Academy but thought that it also had its better moments. "There were some instructors and officers I really admired and still do. It was just important to be able to talk to somebody who'd been through it all. The rifle team was really my salvation, though. From November to June I'd practice every day from 3:00 to 6:00 or 6:30 P.M. It was a time I could get away from barracks; it was kind of an emotional release. Mess hall was a release, too; I figured I could relax, sit down, talk to friends, have a cup of tea, and not worry about anything. And the movies were good. I guess you could say that anything that took you away from the sterility and monotony of the place was good and at least made things tolerable."

There is no escaping the fact, however, that many cadets do resign. The usual four-year attrition rate—thirty-three percent—is much higher at West Point than the drop-out rate at the Ivy League schools (seven percent), but it is still considerably lower than the average for all colleges and universities throughout the country—a staggering sixty percent. But considering the careful selection procedures and relatively high quality of West Point cadets, the attrition figure for the Academy is impressive indeed. Over two-thirds of the cadets who leave resign voluntarily. About a third are forced to resign because of academic difficulties, violations of the honor code, or lack of aptitude for military training.[1] Most of the attrition—twenty percent—takes place in the first year. Al-

56

most eight percent leave in the second year, three percent in the third, and two to three percent in the final year. The proportion of cadets who resign voluntarily (as opposed to those who are forced to leave) is higher among freshmen and sophomores than it is among juniors and seniors, who face military service as enlisted men if they resign after the start of their third academic year.

Studies carried out at West Point by the Office of Institutional Research reveal no major differences between voluntary and involuntary* resignees. In fact, most involuntary resignees leave for precisely the same underlying reasons as many voluntary ones. Only the mode of separation is different. For example, a cadet may feel that flunking out is less of a stigma than resigning, so he neglects his work and is forced to leave for academic reasons. Although the Office of Research was cautious about drawing firm conclusions from the relatively small sample of cadets included in their study, it noted some interesting differences between voluntary and nonvoluntary resignees: the parental income of voluntary leavers is significantly lower than that for nonvoluntary leavers; voluntary leavers tend to come from smaller towns than involuntary ones; the percentage of Catholics is higher among voluntary leavers than among the involuntary; the voluntary resignees are significantly more liberal in their political beliefs than the nonvoluntary. This study also found that only sixteen percent of cadets who resigned regretted their action. With the possible exception of parents in only a few instances, no one person or group, they said, could have influenced them to stay. The majority of them served in the army at some time after resigning from West Point (seven percent became commissioned officers), but ninety-four percent ultimately went on to college elsewhere.[2]

The most common reason cadets give for resigning is "no desire for a military career," a rather nonspecific statement which reflects, I think, their disaffection once they see the reality of Academy life. I suspect also that a lack of desire for a military career overlaps considerably with the next five most important reasons given for leaving West Point: lack of personal freedom, inflexibility of the Academy environment, pettiness of West Point, lack of consideration for the individual, and inequality and/or pressure from the "system." Resigning cadets also mentioned their disillusionment with education at the Academy, the indifference and apathy of upperclassmen, the poor quality of Academy officers, the five-year service obligation they have after graduation, and "the brain-

---

* Even if a cadet is forced out of the Academy, he is usually allowed the official courtesy of resigning.

washing of cadets to accept violence and war." "I'm not a killer," said one cadet who left. Another wrote that since he'd left the Academy, "life seems less serious and I've developed an aversion to shooting at man-shaped objects."

In other words, most cadets resign because they dislike the Academy and the way it treats them. Some also leave, however, because they wish to pursue their studies in an area in which the Academy offers only a few courses. Anthropology, biology, and psychology are but three examples.

While resignations are higher than the Academy would like and for reasons which it might rather not hear, the attrition rate can be put in a different and—from the Academy's standpoint—brighter perspective. A study done at the University of Pittsburgh and published in 1971 revealed that only two out of every ten high school students enter the occupation they chose in high school.[3] Approximately six or seven out of every ten high school students who enter West Point graduate at the end of four years and serve at least five years in the army, although forty percent of seniors in the class of 1971 said they had no intention of staying in the service twenty years. If in fact that forty percent resign and go into another line of work after their five-year commitment is over, this means that approximately four out of the ten students who entered West Point will make the army their career. In other words, forty percent of high school students who enter the Academy make the army their permanent profession, whereas only twenty percent of students at large end up in the profession they chose in high school.

Resignations during Beast Barracks were a somewhat different matter from resignations the rest of the year. A disproportionately large number of cadets—thirty-four percent of the four years' total for any one class— leave voluntarily during the first three months of the year. New cadets have not yet formed the personal ties which older cadets have found so valuable. The shock of encountering a world so different from the civilian world most new cadets have always known is greatest at the beginning. The three most frequent reasons new cadets give for resigning are lack of personal freedom, the Academy's absence of consideration for the individual, and the pettiness of West Point. "Lack of motivation to begin with," lack of maturity, pressure of the system, and the inflexibility of the environment which left them little time for themselves were cited next in order by resigning new cadets.

I had the impression that certain cadets tended to wash out during Beast Barracks earlier than others: those who liked to get off by themselves, those who had a penchant for music or poetry, those with obsessive personalities who liked to concentrate on one thing at a time

(Beast Barracks, with its many demands, was simply devastating for these people), those who could not relate well to others, those who came only to please others or for the prestige of West Point, and, perhaps the largest group, those who had mixed feelings about attending West Point in the first place.

Most parents naturally hated to face the possibility that their son wanted to leave or would be forced to resign from West Point. They invariably encouraged him to stay a bit longer, to wait before making such a big decision. Sometimes their own hopes and aspirations for their son were on the line, of course. They knew that West Point means a free education as well as a full-time career. "My father was a lieutenant in the army during the Second World War," said one cadet, "and of the mind that if you graduated from the Academy you were set for life." It was said among cadets that a father who had himself attended West Point was often more sympathetic to his son's desire to resign than non-West Point fathers were. He knew what it was really like. In the end, though, most parents accepted their son's wish to resign.

There were some notable exceptions. One cadet from South Carolina, the first in his county to be accepted at West Point, came to the Academy in response to excessive pressure from his father. Within two days he regretted his decision. He cried every day, was unable to remember his "poop," and became extremely nervous. When I asked him if he was sure he wanted to leave, he burst out, "Sure? I've never felt so sure about anything in my life." He refused to return to the barracks. "I can't take it anymore; I'd run away first." Unfortunately, his father, who had himself resigned from the Citadel years earlier because it was too difficult, was unsympathetic. If his son decided to leave, he told him in the only letter he ever wrote the boy, he might as well kill himself rather than come home. When his son did in fact return, however, he relented, though grudgingly.

Resignation was a big issue for every cadet during Beast Barracks. Most of them thought of it seriously. The fear of being considered a failure was especially acute during the summer. "If I leave here, how can I ever walk down the street with my father again?" asked one cadet who did in fact resign a month later. Intense pressure was applied to cadets who wanted to resign not only by their parents but also by upperclassmen and tactical officers. One of the commonest techniques they used capitalized on this fear of failure. Cadets were told that Beast Barracks was the supreme test: if they failed in Beast, they would fail in life. Of course this argument, which added to every cadet's anxiety but rarely persuaded him to change his mind, was false.

59

If a cadet did decide to leave, he found it impossible to resign grace-fully. Cadets who left were often called "quitters" or "pussies." Those who stayed were "studs." "When I wanted to go at the end of Beast Barracks, they called me a 'quitter,' " said one cadet, "so I stayed. But now that I'm planning to resign after my second year, my tac officer has called me a 'leech.' He claims I stayed here only to get a free education at the expense of the government."

The period following resignation was never easy for new cadets. One resignee who enrolled at Yale after he left the Academy in August 1970 wrote

To tell the truth, it was a mistake to go to college right after I left West Point. I should have taken time off. I've encountered other students who left the service academies under similar circumstances, and I find that they have had the same problems as I. Leaving the Academy during or after "Beast" seems to cause a state of apathy for a considerably long period afterward. My attitude of complete indifference almost ruined my college career before it got off the ground. Getting good grades, the prestige of Yale, even the three-thousand-dollar scholarship money I was wasting meant nothing to me. I had lost my motivation. I'd heard too many tac officers tell me about being a "star" man, too many speeches about being the cream of the crop. Somehow the lack of interest I acquired about being an "Old Gray" was transferred to any ideas I had about becoming an "Old Blue."

Of course, my observations are hardly broad enough to have any clinical validity, but ask a few people who left during New Cadet Barracks. Ask the boy who has had to go home and face, if only for a short time, the disap-pointment in his mother's eyes, the disappointment mirrored on his father's face. Ask the boy who has walked down the street and met with too many well-meaning friends and neighbors who greet you with "What are you doing here? You're suppose to be at West Point."

Maybe I've overgeneralized, but next New Cadet Barracks, PLEASE tell those who decide not to stay that going there was a big step, but it's a long road back.

In the summer of 1971 an experiment took place at West Point that showed what a great role the environment of Beast Barracks plays in causing resignations and psychiatric casualties. The previous summer had been a psychiatrist's nightmare: thirty-seven psychiatric casualties, including eleven suicide gestures. The boarders' ward, a separate facility to which all resigning cadets were confined while being processed out of West Point, was filled almost to overflowing as the end of August ap-proached. Until 1971, no cadets were allowed to leave during Beast Barracks once resignations had reached or threatened to reach a certain arbitrary number. Those who insisted on resigning were either sent to the boarders' ward or forced to remain in their companies. They were allowed to leave when Beast Barracks was over, and they did, usually in droves. In 1970, for example, eighty-eight cadets were allowed to resign during July and August, a number identical with resignations of the summer before, which lent some credence to the persistent rumor that whether or not the colonel in charge of summer training got a star (a promotion to brigadier general) depended on his resignation rate's being no higher than his predecessor's the summer before. No wonder a new cadet complained, "It's easier to get into West Point than it is to get out." By the end of September 1970, however, an additional eighty-six cadets resigned; although most of these resignations had been held over from Beast Barracks, they were not charged against the commander of summer training.

But in 1971 some dramatic changes took place in Beast Barracks. The officer in charge, a colonel, was directly responsible for the modifications, but much of the information on which he based his decisions came from our psychiatry service and the department of military psychology and leadership. First, cadets and officers were selected for Beast Barracks more carefully than they had been in the past, which meant that a more sensitive group of individuals trained the new cadets. Second, new cadets were treated more humanely. Excessive harrassment was discouraged, cadets were able to eat more easily at mealtimes—a radical departure from a long tradition that meals were prime time for recitation and hazing—and most were able to get more sleep than new cadets usually got in Beast Barracks. The third and most important change was the inauguration of a flexible resignation policy which made it easier for cadets to leave the Academy. The company tactical officer was informed, within forty-eight hours, of any new cadet who wished to resign. This relieved upperclassmen of spending unrealistic amounts of time on cadets who were determined to resign and gave the resignees the assurance that their request had been heard. It also eliminated the awful feeling among new cadets that they were trapped at West Point. Although there was considerable anxiety in the command that resignations would soar out of control with this policy, this did not occur. A substantial increase in resignations occurred by the end of August 1971 as compared to the year before—124 compared to 88—but by the end of September the numbers were almost identical—176 and 174, respectively. The result, in fact, was that resignations at the end of the three-month period were no higher than in the past and individual suffering was certainly less.

61

With these changes the number of suicide gestures dropped to only two, a striking decrease from the eleven of the year before. Other important differences between the two summers were thirty-three percent fewer visits to the emergency room for any reason (which amounted to several hundred fewer visits to the hospital), a smaller number of cadets referred to the psychiatry service, fewer psychiatric hospitalizations, and no discharges at all for psychiatric reasons.

These changes, which a psychiatrist might regard as signs of progress, were not always greeted so warmly by the corps. Some new cadets and upperclassmen felt that an easier Beast Barracks diminished their own achievement. At the end of every summer, a number of new cadets complained that the ordeal had not been difficult enough. Some of this was, of course, bragging, but these cadets also felt genuinely disappointed, as if they had not been tested as rigorously as their predecessors. An army doctor said, "Many new cadets' response to changes in Beast Barracks is negative. They came feeling it would be a test of their manhood. To win membership in the club, they feel they have to be tested; the worth of the experience is directly related to the amount of suffering they have to go through. Many felt that last summer [1973] wasn't tough enough, that too many pussies got through; and they don't want to hobnob with wimps."

These changes could also stigmatize a class permanently. "The class of 1973 was considered a pansy class because they had an easy Beast Barracks and fourth-class year," said one cadet. "They were known as 'wrinkle-free '73' because they were the first class who didn't have to brace the old way." Even West Point officers who felt that the class of 1973 showed less spirit and leadership ability than other classes were certain this state of affairs had something to do with a more lenient Beast Barracks experience. I suspect, however, that this was a case of a self-fulfilling prophecy: having been told they were not so good, not so tough, as previous classes because their Beast Barracks had been a departure from tradition, the class of 1973 might well have taken it to heart, become demoralized, and in fact failed to perform as well as other classes.

Although Academy officers often seemed to ignore the role of the environment while lamenting their drop-out rates (a comment typical of officers was made by Lieutenant General Albert Clark, a West Point graduate until recently superintendent of the Air Force Academy: "The art of selecting winners and losers," he observed, "is not an exact science."), the problem of attrition at the Academy is more than just a problem of interaction between a cadet and his West Point environment. Indeed, it has become a problem for which all the major service academies, not only West Point, wish they could find an answer in light of their soaring resignation rates. The attrition rate for the class of 1974 now stands at forty-two percent for the Air Force Academy, thirty-two percent for Annapolis, thirty-eight percent for West Point. Several hypotheses have been advanced to explain this phenomenon. One is that cadets who came to the Academy to avoid the draft have resigned since the draft ended in early 1973. Another is that cadets are more disaffected with the Academy these days because the contrast between the rigorous life of a West Pointer and the relatively liberal environment of high schools and colleges is so great. A West Point officer mused, "There are no absolute values anymore, no good reasons for going through the kinds of things cadets have to go through here at West Point. There's no longer any nobility in sacrifice." A third holds that unfavorable publicity has finally caught up with the Academy. Its public mask has been rudely pulled off. The recent spate of newspaper articles have exposed hazing practices at both West Point and the Air Force Academy, serious violations of constitutional rights at all service academies, and the cruel procedure known as "silencing" at West Point. As the Academy becomes a focus of debate and criticism, an atmosphere is undoubtedly created which allows cadets to express their own feelings and doubts more freely. At times, cadets' morale is precarious enough. Without reward for their sacrifice in the form of public support and appreciation, some may well become even more discouraged and resign, especially those who came to West Point because of its previous high status in the civilian world.

The Academy has always been anxious to maintain a particular public image, even while hiding the realities of West Point life. A recent study published by their Office of Research found, in fact, that young men presented with a realistic notion of what to expect at West Point are more likely to stay at the Academy than men who are not. Twenty percent of entering cadets in the 1975 class were sent a small pamphlet titled "The Challenge" before they arrived at West Point in 1971. The booklet described Academy life honestly and informed entering cadets of such matters, for example, as the lack of time they would have to themselves and the demanding schedule. Cadets who did not receive the pamphlet had almost twice the resignation rate of the men who did.[4] It will be ironic indeed if it turns out that West Point, which has so possessively guarded its true character in isolation from the civilian world, has actually been doing itself a disservice for many years.

7

A WEST POINT EDUCATION

Education at the Military Academy labors under a double burden of engineering and military training. Both engender attitudes and ways of thinking that are antithetical to education. Education should develop the ability to reflect and think critically. Engineering develops a methodical, orderly, precise mode of thinking that aims at solving concrete problems. Military training demands loyalty and obedience.

Attempts to house education, engineering, and military training under one roof have always caused problems, as West Point's history shows. The Academy was founded in 1802 during Jefferson's presidency as an engineering school and, until 1866, was actually a branch of the corps of engineers. West Point placed no special emphasis on training soldiers during the first fifty years of its existence. Until the Civil War, more Academy graduates became railroad presidents than generals. The first chairmen of the engineering departments at both Harvard and Yale were West Pointers. In 1850 the president of Brown University said that West Point graduates did "more to build up the system of internal improvement in the United States than [graduates of] all other colleges combined." The Tactical Department, whose responsibility is military training, was established in 1858; but not until 1900 was professional military training (as opposed to technical training) really emphasized.

While reducing the technical content of the curriculum, West Point still tried to crowd a liberal arts education and military training into a single course. As Samuel Huntington observed, "The effort to achieve both goals caused continuous tension, a crowding of the curriculum, dissatisfaction on both sides, and persistent suggestions for reform."[1] These difficulties are as great now as they were seventy years ago.

Although West Point graduates were first awarded an academic degree in 1933—a Bachelor of Science—it was not until World War II that the Academy placed heavy emphasis on education for its prospective officers. As the role and size of the army expanded to meet increasing worldwide commitments and as military men penetrated into governmental councils in growing numbers, it became obvious once more that the old combat virtues were not enough. A broad liberal arts education was felt necessary because of the complex and diverse roles an officer assumes throughout his career. He was no longer an individual who passively took orders, the army reasoned, but, instead, a man who must work closely and knowledgeably with foreign affairs experts, educators, scientists, and legislators, a man who could be called upon to prepare legislation, the national budget, and the American position on foreign policy issues. Above all, he was a man possessed of intimate knowledge of the social and political consequences of military action.

West Point makes heroic efforts to produce this well-rounded man. Cadets attend classes four and a half hours a day, five days a week, in addition to three hours of classes on Saturday. Classes begin at 7:45 A.M. and end at 3:15 P.M. Every class is mandatory, and anyone who misses is "awarded" a handsome punishment. Freshmen and sophomores must study in their rooms for three hours every evening, which means they have approximately forty-five minutes to prepare for each class. Juniors and seniors may roam unconfined in the evening, but they still carry the same heavy academic load.

During his four years at West Point, a cadet spends approximately fifty-nine percent of his time on academic subjects, forty-one percent on military subjects. The curriculum is heavily weighted toward science and engineering courses. In terms of total semester hours, each cadet spends forty-eight percent of his time in math, science, and engineering, thirty-nine percent in humanities, and thirteen percent in electives. In the first year alone, seventy-one percent of the time is spent in science courses (forty-three percent in math alone) and twenty-nine percent of the time in the humanities. In his senior year, the cadet devotes seventy-two percent of his academic time to the humanities (excluding electives) within the standard core curriculum.

By far the largest number of semesters—eight—are given over to engineering. English, foreign languages, history, and social science receive four semesters each, followed by math and physics with three apiece. Two semesters are devoted to the study of environment, chemistry, law, and psychology. Electives consume another six to eight semesters. The number of instructors in each department also reflects West Point's engineering bias. Four engineering departments—Earth, Space and Graphic Sciences; Electrical Engineering; Mechanics; and Engineering—have a total of 148 instructors, followed by mathematics with 71, foreign languages with 58, social sciences with 52, and English with 48. The course offerings are diverse and reasonably sophisticated, though, as Ward Just observed in *Military Men,* the most sophisticated courses are offered by the engineering and mathematics departments.

West Point's goal is to produce a generalist, a man, who, according to the catalog, "is prepared to lead the smallest combat unit or to advise the highest governmental council." For this reason, academic majors, which tend to produce specialists, do not exist at West Point. Cadets may "concentrate" their study in one of four areas: basic sciences, applied science and engineering, national security and public affairs, and the humanities, though a quota system is used to assure that no more than ten percent of cadets concentrate in the humanities, no more than thirty-five percent in the other areas. It is no accident that enrollment in the humanities is limited to ten percent of each class. For a profession oriented toward weapons, technology, and foreign policy, English or history has less relevance than engineering, computer science, or a course in Soviet government.

It may be, too, that the study of literature subtly undermines military training. Literature forces one to think in terms of individuals, not abstractions. It becomes difficult to think of a character in a novel as merely an object. Literature also emphasizes the relativity, not the absoluteness, of ideas and values. As one West Point officer said, "For me the study of literature has made it harder to stay in the army. As you read you begin to see all kinds of possibilities; you even begin to believe matters can be settled rationally by discussion. That's not true in the army." The only novel plebes read in its entirety during their freshman year is Mark Twain's *Pudd'nhead Wilson,* a book which, ironically, exposes the foolishness of rigid, categorical thinking.

The Academy may have solid grounds for limiting humanities majors. West Point's younger relative, the Air Force Academy, has long taken pride in its more liberal curriculum, its emphasis on the humanities and social sciences, and its relegation of athletics and even military training to a minor place in the curriculum. The Air Force Academy also has the highest attrition rate of any of the service academies: forty-two percent for the class of 1974. It has been suggested that its liberal approach is the very cause of its problems: a reputation for liberalism may attract students most likely to be disaffected with military life. In 1971, the Military Academy did a quiet study of the records of six cadets who had become conscientious objectors while still at West Point. The social sciences department, which has the reputation of being the most "liberal" department at the Academy, was suspected of promoting this radical change of heart. When the records were scrutinized, however, the only significant denominator was that all six cadets had taken an advanced course in English their freshman year. A study done on civilian students at the University of Pittsburgh and Ohio State revealed that humanities majors, compared with social science, natural science, and engineering majors, were significantly less willing to obey immoral orders or to advocate the use of nuclear weapons in war, were least interested in endorsing Steven Decatur's statement, "My Country, Right or Wrong," and were most critical of the military budget.[2]

From the curriculum alone, one can see that West Point's concept of education is different from that of a civilian college. In the civilian world, the goal of education is the development of understanding and reason. Ideally, this goal is attained through free, critical thinking and reflection under the tutelage of men who themselves are committed to teaching, learning, and research. Knowledge is presented in the form of alternatives rather than as indoctrination. Personal growth, the development of each individual's greatest potential, is an important part of education, requiring openness to new experience, curiosity, relative rather than absolute values based on one's own experience, personal flexibility and creativity.

What I have described is an idealistic concept of education which is, admittedly, rarely attained at even the best liberal arts colleges devoted exclusively to education. The cultivation of an atmosphere in which real education can take place is difficult anywhere. How much more difficult, then, is it in an institution which has other goals besides that of education.

As the Academy's catalog readily informs us: "The academic curriculum and military training encourage logical analysis, clear and concise expression of considered views, and independent thought and action, along with a *readiness, developed within the framework of military discipline, to carry out orders without reservation once a decision has been reached.*" (Emphasis added) This willingness to carry out orders without

67

reservation is a large part of what West Point calls "character," other distinguishing features of which are loyalty, courage, dignity, and respect for rank and tradition. Character-building is explicit in the Academy's academic mission. As Dwight Eisenhower said, "Character is the Academy's concern, a concern to which it has always given single-minded, almost fanatical devotion."[3]

Character-building as an explicit part of academic work is indeed unique, but perhaps the army word "mission" is what really distinguishes military education. Everything in the army, education included, is secondary to the mission. Education is never valued for its own sake, is never an end in itself, but rather a means to an end. Instruction in chemistry, nuclear physics, electronics, and astronautics, for example, was increased at West Point in the last decade because of the "increased technological character of weapons and techniques of war." Similarly, a concern with national security policies led to more instruction in geography, history, government, economics, and international relations. As the late David Boroff, in a series of articles on the service academies for *Harper's* in 1962 and 1963, clearly saw, "The military style, with an emphasis on completing a mission, does not encourage loose ends—and it is the loose ends of one thinker which provide the urgencies of the next. Cadets had a tendency to give final, definitive responses—mission completed, sir!"[4]

Military education is, in fact, hardly education as most civilians think of it. Civilian education benefits the individual; military education benefits the army. Civilian education aims at personal growth; military education aims at socialization and indoctrination. Civilian education encourages a person to define himself; military education defines a person. Civilian education values self-expression and creativity; military education emphasizes control and inhibition. Also, the spirit of military education is different. Studying another country with an interest in its people, its culture, and its civilization is very different from studying another country for reasons of national security. Indeed, the military world has its own word for knowledge: intelligence.

Education at West Point assumes its military colors very quickly for most cadets. It is not until two months after their entrance into the Academy that cadets get their first exposure to academic work. Many officers and cadets are convinced it is ill-fated. A survival mentality, learned during Beast Barracks, carries over into the classroom. An officer in the English department, once a cadet himself, commented: "By the time new cadets come to reorganization week they're very motivated about academic work. But between Labor Day and Christmas, the jig is up. That's the end of academics for most cadets. They watch upperclassmen; it doesn't take them long to see who's the cool guy—the guy who does as little work as possible. Academics turns into another part of the game—you just do what's required to get by. Cadets see that there isn't much correlation between hard work and achievement."

Donald Cantley, a senior who resigned from the Academy in 1972, wrote in *The Pointer,* "Rebounding from Beast with high hopes for an improved future and a worthwhile academic experience, many cadets discover not a learning but a military environment on the first day at the boards in plebe math. They discover a new regulations book—not a guide to writing—in the style manual. And they discover paranoia—not the flowering of their minds—in their French instructor's daily inspection."

The French department at the Academy is notorious among cadets for the zealousness of its inspections. One cadet said, "It's by far the worst. Guys have gotten one demerit for not wearing a collarstay." Another cadet observed, "Tactical officers have a military responsibility; they have to be concerned with how guys look and behave. But the academic departments ought to be primarily concerned with getting things done in the classroom. I've always felt that the responsibility of the academic officer is to teach, to lay off the shoes. When a department tries to go tactical, it loses. The French department is a good example. They've lost people on electives because guys have to come into class, sit down, look straight ahead, not cross their knees, and have their shoes inspected. And when cadets take fewer electives, there's less need for instructors in that department, which, in turn, weakens the department's power. That's what's happened to the French department. Officers who teach Portuguese don't have the same hang-up about discipline, which explains why Portuguese is one of the most popular foreign languages at West Point. Officers have a hard enough time interesting cadets in anything, anyway; discipline in the classroom just turns them off."

The French department is not unique, of course; plebe math, physics, military science also have their periodic shoe inspections. Offenders are "awarded" three demerits for unshined shoes. In this disciplinary atmosphere, preparation for class is also considered a military responsibility. Five to seven demerits are awarded to cadets who attend class "improperly prepared for instruction." In one English seminar, any cadet who fell asleep was forced to stand for the rest of the period. Being intentionally late to class may cost a cadet fifteen/twenty demerits and hours of marching.

Sylvanus Thayer, the superintendent at the Academy from 1817 to 1833, developed a system of education which emphasized stern discipline, a marking system with daily grades, division of classes into small sections on the basis of academic merit, and daily study and recitation. The Academy has modified these procedures somewhat—daily grading is infrequent in the humanities and social science courses, and cadets are no longer required to recite every day—but Thayer would have no difficulty recognizing his system were he to return to West Point after 150 years. Cadets even have a saying: "West Point: 150 years of tradition unhampered by progress." Grades for every course are still posted every week, and sectioning—dividing cadets into different teaching sections every three weeks on the basis of their grades—still occurs. There are two reasons for sectioning. The first is that it helps the instructor avoid favoritism toward any particular cadet. Fairness and uniformity in grading demand a totally impartial evaluation by the instructor. Second, sectioning also allows for individual rates of learning. A man who consistently does poorly in his section will be progressively dropped back until he ends up in a section with men his own speed.

This unending insistence on uniformity is another reason the Academy has been reluctant to institute academic majors. If cadets take different courses in different fields, it will be impossible to compare, grade, and rank them equitably in order of merit. And, needless to say, the demand for uniformity severely inhibits individual expression. In plebe English, for example, cadets are instructed to follow a style manual when they write their compositions. This manual instructs them in the mechanics of writing a paper: where margins should be set, how much space should be left at the top and bottom, where footnotes should be, and so on. But the manual is often used more as a club, a tool of punishment, than as a guide. As one English instructor said, "Style becomes more important than content. You'd be shocked at the number of times I've heard instructors in the department laugh and say, 'Mr. J. [a cadet] wrote a beautiful essay, but I had to give him a D because he didn't follow the style manual.'" As one senior remarked, "The style manual is an obedience thing."

Another feature of West Point life that contributes to uniformity is the "Cooperate and Graduate" syndrome. The cadet who works too hard is criticized; he raises the grading curve and makes it difficult for his classmates. A man who is out front is expected to help his less fortunate classmates rather than proceed at his own speed. If not, he runs the risk of "getting hit"—receiving low aptitude ratings from his classmates. "If a guy can be at the top with no sweat," said one cadet, "fine;

we admire that. But if someone has really worked to get there, he's liable to be unpopular." While this may promote cohesion and camaraderie, it obviously stifles personal initiative.

The relationship between cadets and their teachers at West Point—generally the ultimate in distance between student and teacher—also hampers education. Informality is difficult; rank interferes. A pamphlet published in the social science department cautions its officers: "The instructor should never allow cadets to forget he is an officer," and then reminds them that the academic officers must demand of cadets the same high standards of bearing, dress, and manner as the tactical department does. Cadets, of course, wear uniforms to class and preface their remarks to the instructor with "sir." An officer teaching English said, "One of the things that's missing at West Point is a place to go and sit down and have a Coke with an instructor. There can be lots of intellectual interchange in the classroom, but it's mostly formal, a reflection of the rank thing." And, obviously, sectioning disrupts teacher-student relationships.

About three-quarters of the instructors at West Point are Academy graduates. Each department has about two permanent professors; the rest of the instructors are on three-year tours of duty at the Academy. Nine percent of the men who teach at West Point (including those in the Office of Military Instruction, the Department of Military Psychology and Leadership, and the Office of Physical Education) have Ph.D.'s; most of the rest have master's degrees. Almost all instructors at the Academy are officers whose primary allegiance is to the service, not to learning and teaching.

West Pointers are singled out for future teaching duty while still cadets. After seven or eight years in service they receive an invitation to return—if they are accepted into a graduate program at a civilian university—to West Point. Most officers accept their invitation readily. Besides being known as a good family post, West Point is thought of as one of the necessary assignments for men on their way up the promotion ladder, though during my last year at the Academy a number of academic officers began to wonder whether a teaching assignment at West Point was in fact the step up it once had been. "Teaching at the Academy is not always considered a military job in the army," said one major. An artillery branch chief who visited West Point in 1971 reportedly told his assembled officers that they would be no further behind in branch promotion than "anyone else with a master's degree." In December 1973 a recent West Point graduate said, "That's sad but true. A recent memorandum from the full colonels' promotion board

in Washington identified 'trends' or career patterns of those being promoted to general officer. Ironically, a West Point assignment—especially
in the academic department—was not a major punch on the ol' card.
Instead, those who stayed in the 'army mainstream' [i.e., command
and operations] were considered most promotable."

Most officers are in their early or middle thirties—captains or majors
—by the time they return to teaching duty at West Point. The military
world affects their teaching as profoundly as it does cadets' learning.
Sometimes they are even treated like cadets. In 1970, the superintendent,
Lieutenant General William Knowlton, let it be known that he expected
all officers to abide by the same haircut regulations as cadets. "Does he
expect us to give up our wives, too?" asked a major. The foreign language department told their officers to use the stairs instead of the elevators to reach their offices in Washington Hall. An officer in the
English department said there was little problem with academic freedom
at West Point, but then, referring to superior officers' practice of checking up on their subordinates' teaching, added, "I've only been inspected
in the classroom three times since I've been teaching here."

Whether or not they know it, officers have considerably less academic
freedom—the right to conduct classes without interference, the obligation to teach honestly and objectively, and the security to do both—
than do their civilian colleagues. Officers are warned not to criticize
other activities at the Academy. If an officer plays the devil's advocate
in the classroom discussion, he should let the class know his real position before the hour is over; if cadets criticize the government or the
military, the officer must make sure the cadets' criticism is "constructive"
and supported by facts.

An officer who resigned his commission after his tour of duty at West
Point was finished had returned to teach in one of the humanities departments in 1969. With good grades from West Point, a fine war record,
and an early promotion to major behind him, he had no thoughts of
leaving the service. Then he became an Academy instructor. "My first
assignment, as the junior man in the department, was officer-in-charge
of the coffee detail. I felt I deserved some recognition, but instead I
was treated as a 'junior guy.' I sensed that they didn't expect much
from the 'junior guy.' " By spring term of his first year, this officer began
to have serious reservations about what went on in his department, considered one of the more liberal at the Academy. "At the end of each
month we were supposed to turn in a list of all the journals and novels
we had read in each four-week period," he said. "But I felt that was
childish—I just wrote 'none' on my card. Some officers complied, others
cheated." A good report evidently meant a better efficiency report. Dur-

ing seminars and discussions within the department, this officer learned that a lieutenant colonel would surreptitiously take notes on the quality and quantity of remarks each junior officer made during the meeting. These notes were also presumably used for efficiency reports.

The same officer, impressed by a cadet's work in class, submitted on his own a good performance report on him. Within two days he received a memorandum, longer than the performance report itself, from his superior. The memo criticized the officer's choice of words, his grammar, and his punctuation. "After that, I wondered if it was worth doing any cadet a favor," he said.

An officer in the social science department summed up the problem with his fellow academic officers: "You need competent people to impart knowledge and inspire cadets. The officers who teach them are enthusiastic, but they're basically third-rate. They have no obligation to scholarship and they're not here long enough to develop any teaching expertise."

In addition to the military assaults it must absorb, education at West Point suffers from another problem: lack of time. In the scheduled, rigid West Point atmosphere there is little leisure, little time to reflect, to pull things together, to connect. This is no trivial problem. Time to think is at the heart of real education. Commented a cadet, "It isn't so much that the Academy discourages thinking, it just doesn't encourage or require it. Other things have priority. It's hard to prepare for Saturday classes because of inspection, for example. When you know the tac officer is going to come through the barracks once or maybe twice that morning, your mind's on that instead of classes."

Further, at West Point education must compete on an equal footing with athletics and military training for cadets' time. In December 1970 an incident took place which illustrates the kind of competition education faces at West Point. An Academy colonel serving on the commandant's staff called the dean's office and requested that an exam period be changed because it conflicted with his son's wrestling match. He became "very angry" when a major in the dean's office told him that such a schedule change involving other cadets was impossible.

Not surprisingly, the ultimate result of all these pressures on cadets is apathy: the majority of them become uninterested in academic work of any kind. "With so many demands, academics are always a duty, never a pleasure," said one cadet. "I think the general attitude toward academics," agreed another, "is that they're just something to do and get over with as quickly as possible so you can go back to sleep."

Many officers are concerned about the apathy cadets show toward academic work. Said a 1962 West Point graduate, "Less than five per-

73

cent of the cadets I've taught in the last three years are really interested in academic work. They never take notes because they know that the day before an exam the instructor will give them enough poop to keep them from going down. But it isn't the cadets' fault that they're apathetic. Our discouragement with them only comes when we forget what it was like to have been cadets ourselves. They're hampered by the system; there's simply too much to do. The apathy we see is a symptom of kids who would like to get involved with something on their own terms but can't. Cadets have to account for every minute. Someone should realize by now that education doesn't work that way." Donald Cantley wrote in *The Pointer:* "One desperate and despondent instructor asked me, 'How can I get them interested?' And I didn't have the answer, because by the time he had posed the question—the second semester of yearling year —the pathology was too advanced."

Apathy also breeds mediocrity. "Most cadets," said one, "just want to keep their weekend privileges, which means a 2.3 or a 2.4 [the grading system is based on a 3.0 scale]. You're either 'pro' [proficient] or 'D' [deficient]; anything pro is good. You learn to run at half speed. You don't want to work harder because, if you do, people will expect more of you."

In some courses—for example, the lower sections of second-year physics—instructors drop any pretense of education. "Spec" and "goat"* methods at West Point refer to the rote memorization of formulas in order to get through an exam. "The prof in goat physics," remembered one senior, "would say, 'Just memorize the formulas in the RDP [Reference Data Pamphlet], then plug and chug.' That means just take the numbers on the test questions and plug them into the formula. That way the test turns into a slide rule exam."

While I think it accurate to say that apathy toward academic work is the norm at West Point, not everyone at the Academy is unhappy with his education. Students at the top of their class rarely complain. In upper-division courses, especially elective courses, no sectioning occurs and cadets do come to know their instructors. Said one officer who graduated in the top one percent of his class in 1972, "Once you got into advanced classes, you got treated differently. For the first time officers took a personal interest in cadets."

Yet, for the twenty or thirty instructors at West Point who, in the words of one officer, "feel they have a genuine civilizing mission at the Academy," the military world with its anti-intellectualism must be sorely trying. An officer who has any intellectual commitment undoubt-

---

* A "goat" is a cadet who does poorly in academic work.

edly comes to a *modus vivendi* with the army by subsuming his gifts to the military enterprise. Josiah Bunting, a former army major and Rhodes scholar who taught at West Point from 1969 to 1972, said: "The intellectual (whom I define simply as a person who thinks habitually and deeply about issues) can fulfill himself in the service. He can have a successful career if he's willing to put his gifts into the service of the army. . . . For the intellectual in the army not to turn into a hopeless neurotic or a frustrated bureaucrat he must be basically dedicated and committed to the army."[5]

In his 1962 article for *Harper's,* David Boroff asked: "Is there any intellectual culture at West Point? Or is the very idea inconsistent with the Academy's mission?" His answer: "There are extracurricular activities of a mildly intellectual nature."[6] I would be more negative than Boroff. West Point is pervasively nonintellectual. The life of the mind has no favored status there. In the never-ending struggle between education and training, education is the loser. The Academy is not hostile to education but tolerates it mainly as a necessary embellishment for the well-rounded officer. An exposure to the humanities, for example, is part of the dress uniform. Both cadets and officers talk about learning as something to be conquered in units rather than as a continuous experience with a personal meaning. As a recent Academy graduate said in a sad testimony to West Point education, "West Point teaches you through its academic system how to become an instant expert in any subject, no matter how little you knew about it initially. Through rote memorization, with little or no comprehension, you learn all the facts about any job you're thrown into."

This concept of education, it seems to me, is not likely to change. Of course the Academy will itself seek to enlarge its claims, partly in an understandable effort to present an attractive public image. But even the best-intentioned critic of military education is obliged to reckon with the hard fact that a training school like West Point cannot be very seriously into the business of liberal education. To say that military training is almost the opposite of liberal education is not so much a slur as it is a realistic understanding of the mission of the military school.

There is, however, one overwhelming reason to insist on the place of education at West Point, or, at least, to lament its low estate there. Education is concerned with the development of the intellect. The basic commitment of the intellectual enterprise is an unswerving devotion to the truth. An institution which undermines this enterprise, however indirectly, does more than dampen curiosity and turn cadets away from learning. It fosters an indifference to the truth. And to do that is to undermine the real basis of integrity and honor.

8

PSYCHIATRY AT
WEST POINT

During my time at West Point the Academy was well equipped to handle psychiatric problems. A complete mental health team, composed of two psychiatrists, a psychologist, a social worker, and several enlisted specialists, was always on call. Cadets were not our only responsibility, but their problems always had priority.

Psychiatry at West Point was very much like civilian psychiatric practice. I did see unusual, even bizarre, cases, but their novelty was infinitely heightened by the dour gray military background against which they occurred. Academy psychiatrists, in contrast to their psychiatric colleagues at other army posts, were not overworked. But psychiatric practice was often demanding, especially during Beast Barracks. Fifty years ago, an Academy psychiatrist wrote, "No system can take a group of young men, dress them alike, drill them alike, and grind them through the same machine without breaking a few of them."[1]

Most cadets who came to the hospital during Beast Barracks had problems that anyone could understand: they were anxious, sad, and lonely. Homesick and appalled at the yelling and hazing, they could hardly eat or sleep. One of the first cadets I saw at West Point turned out to be one of the most typical. In early July 1970 a trembling, anxious cadet was brought to the emergency room. Reassurance and a mild sedative quickly calmed him down. The oldest of four children, he was born in Belgium and brought to the United States when he was three. His father, a steelworker, had always wanted one of his sons to attend the Academy. The young man had applied to West Point in the middle of his senior year in high school, but even then he was uncertain whether he really wanted to come. His problem was solved, he thought, when he obtained only an alternate—rather than a principal—appointment to the Academy. He decided to go to the University of Indiana, where, by the time his appointment was announced (after the principal candidate was disqualified for physical reasons), he had reserved a room, paid his fees, and selected a roommate. His father told him he would be an idiot not to go to West Point.

By the third day of Beast Barracks, the cadet knew he had made a bad decision. "I can't stand all the yelling, the constant criticism, and no sleep." Two days later, he notified his squad leader that he wished to resign. But nothing happened. Over the next week, he became more tense and irritable and hardly slept at all. He felt trapped at the Academy and began to worry that he *couldn't* leave. Finally, he fainted at supper one night and was brought to the hospital.

He was hospitalized overnight, then allowed to return to his barracks. In a second interview, on the morning of his discharge, he recalled that he had had similar nervous episodes in high school just before swim meets. This time a good night's sleep had apparently restored his equilibrium and confidence. After returning to his barracks, however, he found that he still wished to resign. He was allowed to leave the Academy two weeks later.

The commonest of the "adjustment reactions" during Beast Barracks was anxiety. This cadet, like a number of his classmates, had mixed feelings about West Point in the first place. Beast Barracks quickly exposed cadets who were ambivalent about coming to the Academy.

But anxiety, although no doubt the commonest problem for cadets during the summer, was not the most dramatic. One desperate cadet was brought to the emergency room after he had threatened to stab himself in the stomach with a bayonet. Another threatened to jump out of his third-story window onto the concrete, and a third reported a persistent impulse to step in front of a moving car. Hysterical conversion symptoms were not uncommon. (In this group of problems, internal turmoil is expressed not through the internal organs—as in psychosomatic disorders—but through a loss of sensation, sight, and the ability to move one's arms or legs.) One cadet was temporarily paralyzed,

77

another lost his voice for a week. A sixth cadet began to jabber incoherently and hear voices that were not there. I often wondered if it was a coincidence that the only cadet I saw in psychiatry during Beast Barracks in 1970 who really wanted to stay at West Point had almost totally lost his hold on reality.

Though I soon became accustomed to dealing with psychiatric emergencies at West Point, I was unprepared at first for what I saw: the casualties were identical to the psychiatric combat casualties I had read about. Those first exhausted, trembling, helpless cadets who were brought to the emergency room created an impression I shall never forget. For the first time I understood the emotional meaning of the term "shell-shocked."

Adolescents experience their feelings—exhilaration, sadness, despair —more intensely than adults. Toward the middle of August 1970, a new cadet was admitted to the hospital for "heat exhaustion." While making my rounds, I caught a glimpse of his face and realized that something other than the heat was bothering him. Over several weeks he gave me a poignant description of the extreme turmoil a person can suffer during Beast.

He was the oldest of seven children, the son of a farmer in Nebraska. He had applied to West Point because "I'd get a free education, free medical care, and all the recognition of being a cadet." Without a scholarship he would have been unable to attend college. His parents were enthusiastic about his decision to go to West Point but insisted the choice was entirely his. He had broken down after running a mile for physical training on the day he was brought to the emergency room. But it soon became clear that the mile run was merely the last straw in a sequence of events which had begun much earlier. Two weeks after he came to West Point, he told his squad leader that he wanted to leave. But then he changed his mind, saying he was not yet ready to make such a big decision. As the weeks went on, he became more discouraged. On the one hand, he could not stand West Point: "I'm trying to live up to an image I don't believe in. I'm losing my identity. I've lost all my self-confidence. They're forcing me to accept values I don't consider important, like saying 'motion pictures' instead of 'movies.' We're treated like children." On the other hand, he felt pressure to stay; his parents and his girl friend would not hear of resignation.

After an overnight stay in the hospital, he returned to his company. His dilemma was by no means at an end, but he was much calmer. I saw him a week later in the clinic. Things were "all right" he said, and no further appointments were scheduled. Another week went by. Then one evening a doctor in the emergency room called me. The cadet was back, crying, anxious, and confused. This time he had collided with another cadet in a football game and had suffered a minor concussion which triggered off the old familiar feelings. "Things are worse than they've ever been. I can't sleep or eat. I cry almost every night. I'm not doing anything for anyone who's been trying to help me. I can't even think straight. I feel like I'm all alone, like I don't have anything to hang onto. I feel like I'm hanging on a rope. I can't see what it's attached to or where it goes. Maybe I'm just holding onto myself."

After this episode, he was determined to leave West Point. "I plan to work for six months, make some money, and get back to school. I want to help people and become a teacher or a poet. I've got to make my own decisions." Yet still he did not resign. He came to the clinic once more, three weeks after the academic year had begun. Nothing had changed. He still disliked the Academy, said he had no interest in academic work, and was absolutely sure he would leave sooner or later. Over the next two years I occasionally saw him from a distance, looking solemn, at a football or basketball game. His tactical officer reassured me that he had adjusted to West Point and was one of the better cadets in his company.

Just before I left the Academy in June 1972 I called him to find out why he had decided to stay after all. To my amazement, he told me that he was on the boarders' ward, which meant he was about to leave the Academy for good. I went to see him immediately. He greeted me cordially and said, "I stayed at West Point for two reasons. There was a lot of pressure from my parents not to resign. They felt I'd throw my life away by doing that. Besides, they have six other kids to raise and they feel responsible for my education, which they couldn't otherwise afford. I also had to show myself that I could take it. Beast Barracks got to me pretty bad and I didn't want to quit when I felt like that. But now I'm leaving. I've proved I can do it if I want. My parents are upset, but that's too bad." Since he resigned before his third year began, he had no obligation to the army. He enrolled at Iowa State the next fall.

Suicide gestures were always a vexing problem at West Point. While the number of completed suicides was much lower than at civilian colleges, the number of attempts—at least within one short period of time— was undoubtedly higher. During 1970–71, eleven of the fifteen suicide attempts which occurred during the year took place during Beast Barracks; nine of these attempts occurred in one month alone. Very few of them reflected a cadet's irrepressible urge to kill himself. A suicide attempt was usually a gesture of desperation, an escape. When a cadet told someone in the chain of command that he wished to resign, he was usually told to "think about it." If he then repeated his request, many days might pass before anything happened. In the meantime, he had to continue with military training. There was no question that this waiting period, so frustrating and uncertain, increased a cadet's despair and contributed to his suicide attempt. In 1970, five cadets had waited over two weeks without knowing whether their request to resign had been acted upon, before they decided to make a suicide gesture. Four cadets had waited at least a week, while two others acted impulsively without mentioning resignation beforehand, though both had been seen by doctors for other reasons within twelve hours of their attempts.

A dramatic example of an individual who displayed a number of medical symptoms ending in a suicide attempt was an eighteen-year-old from the Southwest who was referred to the psychiatry service by an ophthalmologist who had been treating him, unsuccessfully, for loss of vision in his left eye. The cadet had never suffered poor vision before, and the eye specialist was baffled. In my first meeting with the cadet, I discovered that he had planned to go to West Point for several years, that he was a good student and athlete in high school, and that he liked the Academy. He was religious, but that apparently did not present problems. He recalled that he had wanted to resign from West Point the first week of Beast Barracks: "It was a lot tougher here than I'd thought." But his parents dissuaded him, and the issue did not come up again. During July he had twice been named the outstanding cadet in his squad. After our first interview, I knew I had not been of much help to the ophthalmologist. That evening, however, one of the West Point chaplains told me that the cadet was a member of a very fundamentalist church. From his experience, the chaplain thought that might have something to do with his problems. The young man's tactical officer reported that he had been teased, though not excessively, for his religiosity.

A second interview with the cadet was as unproductive as the first. Two days went by. His vision did not improve, and he looked increasingly unhappy. With his permission, I undertook an "amytal interview." (Sodium amytal, used intravenously, lessens a person's inhibitions about talking. It is popularly known as "truth serum.")

The results were unexpected and startling. As the drug began to take effect, the cadet started to cry and said, "I only feel sad when I think of Christ dying for our sins, especially mine." He went on to say that, three years before, two high school classmates had been killed in an automobile accident. He had ignored them before the tragedy. "I was arrogant." Had he been able to "witness" before them (i.e., profess his belief in Jesus Christ) he believed he would have saved them. In the following months, he had had several dreams about the accident, dreams which inspired him to spread "The Word." Even though he was ridiculed "ninety-nine percent of the time" by classmates, he was encouraged by his parents and church members to carry on.

He had come to West Point determined to "witness to the corps," a project he figured would take him about a year. Unfortunately, he was taunted unmercifully by his classmates, who quickly gave him the nickname of "The Preacher." On the day he lost his sight, he had been ordered to stand in front of his squad and deliver a sermon. He considered this a good opportunity but was ridiculed as usual. Also, in this hostile environment, he had begun to feel guilty: "The pressures are getting to me. I feel I'm losing my faith."

The other horn of his dilemma was that he had worked hard to come to West Point. His small town had given him a splendid send-off when he left. His parents urged him to stay. He felt trapped. He could neither remain at West Point nor return home.

It was clear that his visual problem was a symptom of this conflict. He could not "see" which way to turn. Moreover, the (unconscious) choice of a medical symptom—near-blindness in one eye—was a workable solution to his problem: he would not be able to stay at West Point if he could not meet minimum physical standards, yet he could return home with some shred of honor. He would not have quit the Academy, but instead would have been medically disqualified.

When I first saw him, the cadet insisted on staying at the Academy. He was therefore discharged from the hospital a week after admission. His vision was still poor, but he could get around without trouble.

Two evenings later, he was brought back to the hospital, severely anxious. He had fainted twice in barracks. After talking with me, he felt well enough to return to his room. The next morning he called his parents and told them he wished to resign. They were sympathetic, though they advised him to think about his decision once more before leaving. He seemed relieved over his decision.

But the next day he showed up at the clinic two hours before his scheduled appointment, looking wan and unhappy. The night before, he had experienced a paroxysm of self-deprecation. "I feel like a total failure," he said. That night he wrote a suicide note, sealed the envelope, took his "medicine," and went to sleep. The note, which was addressed to me, went:

It is with regret that I write this note, but it is something that I felt I must do. So you'll know what I took, I took six Bufferin out of my dop kit, two aspirin, and a small bottle of Scope. At least I'll have a clean mouth. Ha ha.

Tell Mother, Daddy, and Joan I love them, and tell them this will keep them from having to claim a coward for a son and a brother. I just couldn't stand having to go back to the company even though my worst conflict is now resolved because of the hassle.

I must thank you for your understanding and help. Thank Dr. M. and all the hospital staff for their kind care. I know you'll think I could have stood it for at least another week, but I can't even stand it now. So I guess I'll "take my medicine" and wait to meet God.

Just try to understand I had to do this. Please do not tell anyone at all about this if it is possible.

Fortunately, the cadet did not succumb to mouthwash and aspirin during the night. He awoke in the morning, groggy, and with no memory of the events of the night before. When he noticed the sealed envelope on his nightstand, he opened it, read it, and showed it to an upperclassman, who brought him to the hospital immediately.

If it had not been obvious before, it was now clear that this young man could not cope with the Academy. The way his symptoms had accelerated, starting with loss of vision in his left eye, proceeding to fainting spells, and ending with a suicide gesture, was clear confirmation. When I notified his parents of the most recent misadventure, they promptly agreed to his resignation. I encouraged him to seek out an ophthalmologist and a psychiatrist in his hometown.

Three months later, in November, I telephoned him. He was working, going to school full-time, living at home, and going to church regularly. But his vision had not improved, a cause for some concern, since follow-up studies of this condition have shown that a significant number of individuals do not regain their vision even over a long period of time. Four months later, however, I received the following letter:

You requested some time ago that I let you know how I was getting along in civilian life. I'm doing just fine, being very busy with school and work. Last semester I received a 3.5 average in fifteen hours of engineering. At midterm this semester, I have a 4.0 going in sixteen hours of work. I am also working in a department store as head of one of the departments. I am dating regularly, although I am not going steady with anyone. I have been bowling at least once a week and playing a lot of pool and snooker. I have joined Kiwanis and am now the youngest Kiwanian in the state.

You will be happy to know that my eye has cleared up completely. The conflicts that I had concerning Christ in my life have been resolved also. I realize that Christ is my redeemer and refuge, and although He is the same for all others, He can deal with them without my interference. I have enough peace and self-assurance that I recently went to the army recruiting office, received a 1-A military status, and am thinking about joining the plain old Army for four years in the next year or two.*

I wish to take this opportunity to express my gratitude for the help you gave me while I was at West Point and for the concern you spent on my behalf. Most of my basic training at USMA is vivid in my mind and all the mistakes are just as vivid, but I have been able to learn something from both.

Although few suicide gestures were serious in intent, they were treated seriously, as even minor attempts could have major consequences. But first, two obstacles had to be overcome. Officers and cadets responsible for training had to be reminded constantly that any suicide attempt was potentially lethal. After several had occurred, officers tended to discount them. "He's just trying to manipulate his way out of the Academy," said one hard-boiled infantry officer after he was told that one of his cadets had made a suicide attempt. Second, it was imperative to make people understand that a suicide gesture did not necessarily represent a psychiatric disorder, that unhappiness and desperation were not the same as mental illness. "He must be crazy to do something like that," was a common response among tactical officers when told about a suicide attempt. This was nonsense, of course, and only allowed those in charge to blame the cadet for any difficulties, rather than indicting the system itself.

At the time that I came to the Point, a cadet who "broke" under the stress of training became the exclusive responsibility of the psychiatrist. But once I realized the motivations behind the suicide attempts, I refused to take exclusive responsibility for any cadet and would not keep him in hospital if his principal problem was that he wanted to resign and was not allowed to. Upperclassmen and tactical officers were thereafter included in all decisions involving a new cadet's future. If a man no longer needed hospitalization and his life was not in danger, where should he go? I asked. Should his request to resign be taken seriously or ignored a bit longer? Sometimes I had to remind officers that if they insisted on keeping a man at the Academy against his wishes, he might make a second suicide attempt.

Psychosomatic symptoms—vague stomach and intestinal upset, headaches, fainting spells, and so on—were also more frequent during the summer than at any other time. If fact, they were twice as common in

---

* Resignees often spoke of joining the army at some later time. This reflected, I think, a desire to compensate for their failure at West Point.

July and August as they were the remaining ten months of the year. Psychosomatic symptoms were always medical expressions of difficulties in adapting to West Point. "The sorrow which has no vent in tears may make other organs weep," a psychiatrist once wrote. In American culture it is more acceptable to have a medical than an emotional problem. Psychosomatic symptoms and their cousins, hysterical conversion symptoms, were often used as unconscious tickets of admission to hospital. It became axiomatic that if a cadet remained in hospital a week or more with persistent intestinal symptoms for which no cause could be found, he had a conflict about West Point. It was also undoubtedly true that cadets sometimes came to hospital to escape training for a short time. The first full-time psychiatrist ever assigned to West Point noted, in 1922, "The medical department is the one flexible link in the military chain: the doctor is the custodian of the keys that open the door to escape; the hospital is the safety valve. The schedule is so rigorous that the cadet is only too glad to consult the physician upon the slightest pretext, in the hope that he may be excused from drill or be admitted to the hospital for a few days' rest."[2]

Most of the therapy I did during the summer was very brief—two or three visits. My first task was to find out what was going on, by talking to the troubled cadet and other people who knew him. I then had to make a decision whether to hospitalize him or see him in the clinic. Most hospitalizations lasted only overnight. We quickly discovered that it was almost impossible to get a cadet back into his training program if we kept him any longer. His symptoms provided him with a big payoff by allowing him to escape the demanding schedule, and his desire to stay at West Point seemed to evaporate. A most remarkable phenomenon often occurred among hospitalized cadets if their discharge failed to take place as planned. If a cadet was due to be discharged at twelve noon, for example, he would usually be all right if he left the hospital exactly when he was told to. But sometimes there would be a hitch: the doctor would forget to write the order, a nurse would not see it, or no one would come to pick up the cadet at the appointed time. Within half an hour, the old symptoms which originally brought him to hospital would return, more intense than ever. It was then harder to get him to leave hospital, and, once out, there was a greater likelihood he would return.

When cadets talked about resigning, I often felt in a dilemma myself. I certainly did not know what was best for many of them. I felt they had to make that decision themselves, though it was often heart-breaking to watch them struggle as they did. My job was never made easier by

Academy officers, who, with that peculiar certainty army people show, assumed that every cadet should stay. For some high-ranking officers, in fact, the worth of psychotherapy was measured by whether a cadet remained or left. During my first summer at West Point, I telephoned the Beast Barracks commander, a colonel, to tell him about a cadet who was having great problems adjusting. The colonel listened impatiently and said, "You must be doing some good over there; he's decided to stay."

Although Beast Barracks contributed disproportionately and dramatically to the total year's psychiatric casualties—a third of all cadets who visited the clinic came during the summer, as did two-thirds of all psychosomatic disorders, a third of adjustment reactions (anxiety or depression), and four-fifths of the suicide gestures—cadets were referred for psychiatric help the rest of the year, too. By and large, the cases were less dramatic and probably more like cases psychiatrists see at any civilian student mental health center. In fact, the more serious psychiatric problems—psychoses and serious depressions leading to suicide—were less common at West Point than they were at civilian colleges. For every thousand students on a civilian campus, a psychiatrist expects that at least one will experience a serious break with reality, and that one out every ten thousand will commit suicide. Between 1965 and 1972, only one cadet committed suicide, and he was at home, not at West Point, when he shot himself. While I was at the Academy, only two cadets were diagnosed as having mental illnesses serious enough to disqualify them from the Academy. It is probably fair to say that the true incidence of mental illness is lower at West Point than in civilian colleges. During my tour of duty at the Point, we saw about one hundred cadets a year. This does not mean, however, that cadets do not have problems. Some of them undoubtedly went to the chaplain or another officer for help. But in the cloistered West Point environment it was unlikely that any serious psychiatric problems escaped detection. Any cadet whose behavior deviated from what was considered normal was quickly sent to a psychiatrist.

Like students everywhere, cadets did not usually require long-term therapy; three or four visits were usually enough. Also, the psychiatry clinic was not a popular spot on post. Only about fifteen percent of cadets we saw came for help on their own; the rest were referred to us by cadet counselors, tactical officers, or instructors. In contrast, over forty percent of students at the University of Indiana seek psychiatric help on their own,[3] while at Harvard the figure is close to eighty percent. Military men are a conservative lot, and seeking psychiatric help is not part of the military ethic. Cadets also worried that a psychiatric visit would go into their permanent Academy file, although this was not the case.

It was possible to undertake more prolonged therapy with any cadet who needed it; the Academy never interfered. Cadets spoke freely and were usually happy to find an officer with whom they could talk.

The majority of cadets we saw each year came from the freshman and sophomore classes. Generally, plebes were sent because they were anxious or depressed, while upperclassmen more often came for evaluation of such diverse problems as sleep walking, poor military aptitude, trouble getting along with other cadets, drinking, and—once—masturbation. I was impressed that, except during Beast Barracks, very few cadets have the identity problems one commonly sees on civilian campuses. Cadets did not ask, "Who am I?" This might have been because cadets had already formed a solid identity by the time they arrived at West Point, but I think it was more likely that the Academy polarized people very quickly: if you couldn't take on any part of a military identity, you simply had to resign. Those who stayed were able to incorporate aspects of the military world into themselves and, in the process, form clear definitions of themselves. Those who continued to dislike the Academy but remained at least established a clear self-concept of what they were not.

Premature ejaculation and impotence were almost the only sexual problems among cadets for which I was consulted, and they amounted to only a handful of cases. Cadets were often sexually naive, probably more so than their civilian counterparts. As for sex education, the course at West Point consisted of one or two lectures a year on "sexual hygiene"—"a title which always makes me think of a man washing his genitals after contact," commented an officer at the Academy. These talks, complete with color slides shown immediately after the noon meal, were given to cadets a few days prior to their vacation and focused almost entirely on the horrors of syphilis and gonorrhea. Rather than instructional periods, these sessions became warnings of the consequences of misbehavior while cadets were on their own.

In the hypermasculine West Point world, homosexuality was absolutely taboo. Cadets rarely even joked about it—an indication, I think, of how forbidden the topic really was. Once or twice I heard vague rumors of homosexuality within the corps, but I treated only one cadet for it during my two years at the Academy. He had a long history of homosexual encounters which had started in grade school, and though he had never approached another cadet, he was worried that he might. I am sure his reasons for attending the Academy were infinitely complicated and largely unconscious. He was no doubt uncomfortable with his sexual identity and hoped that by going to the Academy, he would prove to himself, and perhaps to others, that he was really a man. And one certainly wonders if the all-male population at West Point held a particular attraction for him. At any rate, this poor man was in anguish by the time he finally came to me. His urge to kiss or touch his roommate—an extremely self-reliant, independent, athletic individual—had become almost unmanageable. He felt his career would be over if he gave in to his desire, and he was terrified that he might do so.

I agreed to treat him. After several months of therapy, his attraction to men had diminished slightly, but he was still worried. Unfortunately, I left West Point at that time and do no know what happened to him. I encouraged him to continue treatment with another psychiatrist at the Academy.

Although I never knew of any documented homosexuality within the corps, this does not mean it never occurs. Any exclusively male community characterized by enforced closeness is necessarily permeated by deep undercurrents of emotional involvement and sometimes by sexual feelings which cannot be easily expressed. But any homosexual activity that did go on would necessarily have been totally clandestine.

But sexual feelings among cadets did come out when defenses were down. Some of the horseplay in the barracks at night—a cadet pulling the hair on a roommate's leg or yanking the blanket off a friend lying almost naked in bed—carried sexual overtones. At a drinking party one night, a cadet allegedly kissed another cadet on the mouth, an act which led to a full-scale investigation by the Academy's Criminal Investigation Division. The young man charged with this indiscretion was allowed to stay at the Academy but was awarded a huge "punishment tour."

No problems caused so much furor at West Point as sexual problems. One cadet was dismissed from the Academy after the C.I.D. discovered that he was sending (in their words) "Polaroid photos of a penis" to girls whose addresses he collected from the missent mail in the cadet orderly room. This unfortunate boy, who had a fine record at the Academy but who was painfully shy with women, took his Polaroid camera to New York City about every two weeks, where, with the aid of a self-timer on his camera and in the seclusion of a rented hotel room, he took pictures of his genitals which he then distributed to young women by mail from West Point.

A psychoanalyst might contend that the young man was only displaying his masculinity, his power, in a primitive way little different in kind from the decidedly more sophisticated fashion of other West Pointers, but the Academy (and the girls) was understandably not amused. The C.I.D. broke the case by identifying the decor of the room in the background of the pictures and finding out which cadet had stayed in the hotel.

Another *cause célèbre* occurred in the summer of 1971 when a cadet on Advanced Orientation Training at an army post in the Northwest was caught inside a large department store after closing hours. The cadet claimed he had fallen asleep in the bathroom, but unfortunately he was carrying a suitcase filled with panties, two bras, a nightgown, two mink stoles, and a couple of wigs. Though he said he had no urge to be a woman, he did later admit that he had been dressing in women's clothes since age eleven. He remembered wrapping himself in his mother's mink coat as early as age six, and later dressing himself in her clothes when he was lonely. Amidst great secrecy and embarrassment, he was flown to Walter Reed Hospital in Washington, D.C., and subjected to several weeks of behavior therapy. Problems began when the well-meaning but naive psychiatrists at Walter Reed tactlessly told the Academy they felt the man had been cured and was not only fit for duty but could also be commissioned. West Point would have none of it. When the cadet was returned from Washington, he was immediately isolated and enjoined from attending classes with other cadets. After a month he was forced out of the Academy permanently.

At about the same time, a recent West Point graduate stationed in Washington, D.C., admitted to homosexuality. This was too much for officialdom. The superintendent at West Point, General Knowlton, received a letter from General William Westmoreland, Chief of Staff of the Army, requesting that the Academy look into the feasibility of screening all seniors for sexual deviation. I wrote a detailed reply to General Knowlton, through the hospital commander, explaining that: a) it was usually impossible to identify a sexual deviant; b) to give each senior cadet a one-hour screening interview would exhaust the resources of the psychiatric service; and c) it was self-defeating, for if a cadet knew he would be expelled from the Academy for a sexual aberration, he would be loath to admit it and even more reluctant to seek psychiatric help. A screening program for sexual deviants was never established at West Point.

In a world that values appearances and order so highly, sexual perversion is the ultimate violation. Fortunately, sexual disorders were rare, though they elicited highly emotional reactions at West Point because they so flagrantly challenged the masculine image so cherished by the military world. Sexual deviation also detracted from the uniformity of the corps, as did all psychiatric disorders. I sometimes wondered if the Academy was not a little embarrassed to have a psychiatrist on post, a man who was not a symbol of order and uniformity but a reminder of the potential disorders and personal idiosyncrasies that lurk within every uniformed body.

I suppose it was this feeling of embarrassment, perhaps mistrust, which West Point officers had transmitted to my predecessor at the Academy. Discouraged, he said to me one night, "The Academy doesn't like to admit we're here. I guess they like to have us around when problems they can't handle come up, but the rest of the time I get the feeling we're pretty dispensable."

Perhaps his reading of the military attitude toward psychiatrists was accurate, but I could not share his discouragement about a psychiatrist's role at West Point, especially where cadets were involved. Of course every military psychiatrist's responsibility is to support the mission by helping men return to productive military duty, and this was done whenever possible. Often, however, it was not possible. By the time I saw many of the cadets, either they were convinced they wanted to resign (and sometimes went to desperate lengths to demonstrate their seriousness), or the Academy had already decided the issue for them. It was vital for these men to have someone with whom they could talk in a friendly, sympathetic, and objective manner, someone who could assure them both that they were not failures if they left West Point and that it was acceptable, indeed human, to have problems.

Cadets were gratifying to work with as well. Their problems, often brought about by the unique stresses of West Point, usually responded quickly and completely to relatively simple treatment methods. Those who resigned, sometimes with an enhanced sense of self-awareness, almost invariably went on to college elsewhere, and I expect that most will make successful civilian careers. The assets that made them attractive candidates to the Academy in the first place are also considerable assets in the civilian world. In most cases, the transition back to civilian life was easier than the cadets had anticipated, though their memories of the military world of West Point will no doubt always be vividly etched in their minds.

9

ETHICS AND HONOR

West Point has long taken inordinate pride in its honor code, the "cherished possession of the corps of cadets and the Long Gray Line of graduates." This is shrewd advertising. The Academy recognizes that young men are intensely, even romantically, idealistic. The prospect of devoting oneself to a cause larger than oneself, of sacrificing at least one's possessions and perhaps one's identity in the service of three glimmering ideals buttressed by an absolute standard of conduct can be powerful inducements to apply to West Point. The honor code, which states simply, "A cadet will not lie, cheat, steal" ("or tolerate those who do"—the mischievous toleration clause), even seems to be an anodyne to the frustrating search for absolutes that marks normal adolescence.

The honor system is run by the cadets themselves. A man who either turns himself in or reports another for lying, cheating, or stealing first notifies the honor representative in his company, who in turn notifies the honor committee. This body undertakes a preliminary investigation of the matter. "It's kind of like a grand jury," said a senior. If this committee decides a case exists, the cadet must appear before an honor board composed of twelve seniors. The decision of the board must be unanimous. If the vote is 11–1 or 10–2, the case is thrown out. If the board decides the cadet is guilty, he has two choices: resign, or take his case to a board of officers. Cadets are discouraged from appealing their cases to the higher board, however. If they resign without protest, no mention of their expulsion will be entered on their record. If a cadet makes an appeal to the officers' board and loses, the Academy notes on his permanent military file that he was separated. A cadet understandably fears that another college will be reluctant to accept him if it learns he has been expelled because of an honor code violation.

If, on appeal, the cadet is found guilty by the board of officers, he is out. But even if he is found innocent, he is not exonerated. His case is reviewed once more by the honor board. If it decides, perhaps on the basis of new evidence, that the cadet is not guilty, the case is dismissed. This rarely happens, however, and until 1973, the usual fate of a cadet in this situation was a form of social ostracism known as "silencing," a virtual sentence to solitary confinement as long as the cadet remained at the Academy.

The procedure worked this way. The disagreement between the two boards was announced to the corps of cadets, who were then asked to vote whether or not to silence a man who had been found guilty by the honor board. A vote in favor meant support of the honor board's decision. In almost every instance, cadets voted to stand by their honor board. But even when the majority of cadets voted for silence, it was impossible to prevent those who voted against it from talking to the person found guilty. While this made silencing less total than the public thought, there was no denying that it was a severe trial and a serious stigma for any cadet. Life at West Point was difficult enough without being forced to live apart from other cadets, walk to meals and class alone, sit at a separate table in the mess hall, and remain incommunicado with the majority of the corps. It was also understood that the stigma would remain with the cadet after he graduated; no West Pointer would talk to him even after he became an officer. Most of the cadets who were subjected to the silence resigned.

A notable and noted exception was James J. Pelosi, a cadet who graduated in June 1973 after nineteen months of silencing. He had allegedly cheated by concealing a pencil lead under his fingernail during an exam in electricity. At the end of the exam, cadets received green pencils with which to grade their own papers as slides containing the correct answers were flashed on the screen. A classmate observed Pelosi writing numbers on his paper with the pencil lead instead of the green marking pencil and reported the incidence to his instructor. Most cadets felt that Pelosi had cheated—"That was never disputed," said one—

91

although Pelosi maintained his innocence and felt that resignation would be tantamount to an admission of guilt. His case was dismissed and he was returned to the corps by the superintendent, General Knowlton, after it was learned that a regimental officer had violated normal judicial procedures by urging the honor committee to expedite Pelosi's case because it was an "open and shut" honor violation.

The Pelosi case provides an unusually revealing picture of the internal dynamics of West Point. "We voted to silence Pelosi," commented a cadet, "because he didn't meet the standards set by the cadets." "Set by the cadets" is the crucial phrase. For, as many cadets agree, the fact is that less than ten percent of honor code violations are reported. The reason? The intense loyalty cadets have toward each other, loyalties far stronger than their allegiance to the honor code. "If you live with a guy for three years, you can't turn him in on honor," said a senior. Another senior said, "I'm loyal to my friends. If it looks like someone won't make it back after taps, I'll cover for him. I know he'll do the same for me. If someone needs help with a math problem [helping a friend with problems outside of class can be an honor code violation at West Point], I'll give it to him. Most of us know where our loyalties are, and we ignore the toleration clause. I don't like to squeal on my brothers."

In light of this, one wonders why honor code violations are reported at all. One reason is obvious: not all cadets tolerate infractions; those who do not are known as "straight" by other cadets and are avoided. Second, in the words of a cadet, "The honor code is an easy way to get someone out of the corps. There are just some guys with bad personalities who're too different, who just don't fit in. They're the ones who get reported." But by far the most important reason, as one cadet explained, is that "Most cadets feel it's unfair to cheat in athletics or academic work; they won't put up with it if they think it affects them directly. Cheating in class raises the curve unfairly. But if a cadet does something that doesn't affect anyone but himself—the cadet who lies to his tactical officer about having swept the floor or the cadet who returns late from leave and says he came back ten minutes earlier than he actually did—he usually won't be reported." This does not mean that cheating in class does not occur; it does, though it is much less frequent than cheating or lying outside class. But cheating in the classroom is risky and will usually be reported if someone observes it.

The norm, then, among cadets is that some cheating and lying are tolerated, but cheating at sports or in class, striving for an unfair advantage in the highly competitive West Point world, is strictly taboo. The majority of cadets will punish such a transgression to the limit of their ability.

This attitude explains the harassment Pelosi was reported to have received while he was still at the Academy. Commented one cadet, "We didn't want anything to do with him and were determined to make it as hard on him as possible. We wanted him out." Pelosi's mail was ripped up; glass was placed under the tires of his car; an anonymous caller threatened to cut off his finger if he wore his class ring.

Resentment of another kind toward Pelosi also surfaced. "If what happened to Pelosi had happened to me," complained one cadet, "I would be out by now. His father's a banker, I think, who was able to hire a fleet of lawyers who threatened to sue the Academy unless Pelosi was reinstated."

Another cadet remarked, "Pelosi was reinstated by our superiors, people above us. They have all the power and can keep him in if they want."

Pelosi thus found himself squarely in the center of a much larger issue at the Academy: the antipathy cadets feel for officers. "Cadets can't identify with their officers, especially those in the tactical department or OPE [Office of Physical Education]," said one cadet. "They're always coming down on us. Cadets and officers are like two enemy camps." Largely powerless themselves, cadets cling all the more tightly to the one small island of independence and power they possess: the honor system. As their handbook says, "For its success, the honor code depends upon the corps, which, individually and collectively, is the guardian of its honor code." An officers' board decision which contravenes their own is regarded as an unacceptable intrusion into cadet affairs. This is the reason the cadet honor board rarely reconsiders its decisions when an officers' board comes to a conclusion different from their own about an honor case.

What, I wondered, would have happened if General Knowlton had ordered the corps not to silence Pelosi? "He wouldn't do it," replied a recent graduate. "First, he'd get flak from his own colleagues for breaking an Academy tradition by interfering with the cadet honor code. Second, the feeling about Pelosi was so strong that he'd have been silenced anyway; and he would have been even more disliked than he was because the 'Supe' was on his side."

A very basic and important difference betwen the cadet honor board and the officers' board complicates every honor case that is appealed. The officers' board follows the Universal Code of Military Justice (the

UCMJ), while the honor board does not. This means that the officers adhere firmly to the concept of due process, while cadets do not. Cadets learn, through their honor board proceedings, that such guarantees as the right to counsel and the right to remain silent are mere technicalities. The honor board does not adhere to the rules of evidence; a cadet may be expelled on the word of another cadet or officer, for example. These proceedings make a travesty of an individual's rights.

Once reported, honor code violations are prosecuted in an inflexible and absolute manner. No distinctions are made among violations. For example, one new cadet was forced to resign for quibbling about his status as a nonvirgin.[1] Another new cadet was expelled from the Academy in 1970 after he confessed he had shined his shoes the night before inspection, not four hours before as he had told his squad leader. A third cadet had to leave because he told cadets he owned a jaguar. A fourth was "found" after he told some other cadets he had no cookies in his cookie box when in fact he did have.

Although many of the fifty or sixty honor cases heard each year are more substantive—one was caught stealing another's camera, another cadet cashed bad checks—most cadets agree that relatively minor, even trivial, offenses that lead to expulsion are all too common. A senior said, "The honor board is blind. All they ask is 'Did he lie?' It doesn't matter whether it's lying about a peanut or a million dollars."

In 1972, several cadets on the honor committee proposed that the honor board institute a "reinstatement clause" into its proceedings. This was an attempt to soften the rigid application of the code; cadets "found" on honor for trivial offenses—those who lied about their shoes or whether or not they had cookies in their cookie jars, and even those who turned themselves in—could be reinstated in the corps. "It was never implemented," said one cadet. "Many guys felt they'd be undermining the honor code even more than it has been. A number of honor cases so far this year [1973–1974] have been thrown out by 11–1 votes. The honor board chairman has threatened to resign because he feels that some of the cadets on the board are subverting it. Besides, how do you discriminate between lies?"

Though the reinstatement clause failed, one outwardly important change in the honor system occurred in September 1973, when the practice of silencing was officially ended by the cadet honor committee. Perhaps the most important reason was the publicity Pelosi's case received. "I think yellow journalism—pressure from *The New York Times* and other papers—had a lot to do with it," said one man. "West Point is good copy these days." As the same cadet also explained, "There were evidently legal problems. The honor system is run by cadets, but it's obvious the Academy supports it. A cadet couldn't be moved out of the company and given a separate table at the mess hall if it weren't for the administration."

This cadet admitted, however, that informal silencing still exists, and that cadets would treat Pelosi today much the same as they did in 1972 and 1973. "Most cadets have better things to do than hassle someone, but of course some cadets would chastise him," said one man. "They wouldn't talk to him, and others would make sure he got slugged whenever possible." A lieutenant colonel observed, "There's no doubt that the articles in the *Times* helped change the policy. The Academy was furious at the Pelosi case and the bad publicity it brought the Point. But it may turn out that the change is really a subtle loyalty test: if a cadet doesn't agree to silence another cadet, it means he doesn't respect the traditions of the Academy."

Honor at West Point may have a narrow definition in theory, but in practice it is far from clearly defined. One of the problems is the blurred distinction between the honor code and the regulations. Cadets who violate the regulations receive demerits, punitive marching assignments, confinement, or all three, while cadets who violate the honor code are expelled from the Academy. The honor code insists on absolute integrity, while the regulations more tolerantly allow for relative integrity. There are some acts which, though they violate regulations, are clearly pranks and nothing more: practical jokes on roommates, or the ceremony of discoloring the private parts of the statue of Washington's horse. There is other behavior, however, that is not so easily labeled. What should one call drinking in barracks, going out after taps, getting out of mandatory parades, or coming to sick call at the hospital to avoid a class? Are these examples of cheating—that is, do they violate the honor code —or are they merely violations of the "regs"? The same problem arises with lying. If a cadet purchases an item in the cadet store, it is understood that he has bought the article only for himself or as a bona fide gift for someone else, as required by the regulations. If, instead, he purchased the item for someone else at the person's request—usually an officer or friend—the cadet would seem to be lying if he states the purchase is a gift for his own use. But, in fact, the offense is not called cheating: it is simply considered a violation of the regulations. The cadet will be punished but will not risk expulsion.

An obvious problem arises here. Cadets, under threat of expulsion, must never lie or cheat. Yet perhaps because of the very seriousness of an honor code violation, most cases of lying and cheating, as illustrated by the examples above, are called by those terms only in the most flagrant cases. They are usually called, instead, "violations of the regulations." One small allowance for lying is made. When new cadets are indoctrinated into the honor code in the first days of Beast Barracks, they receive instruction in "social honor": it is acceptable to lie in all social situations in order to avoid hurting another person's feelings. "Suppose an officer invites you to dinner and his wife cooks a bad meal," explained one cadet. "If she asks you how the food was, you're expected to say, 'Very good, Ma'am.'" (Until one cadet was expelled for lying after he told a girl who asked him for a date that he was "cold beverage corporal" that weekend, lying to avoid hurting a girl's feelings was also considered "social honor" by some cadets.) Of course the Academy is right to make this allowance. To apply the honor code to a cadet's every action would be priggish and unrealistic. Simple tact makes social honor imperative.

But the two different standards of conduct prescribed by the honor code on the one hand and the regulations on the other foster a hermetic view of morality. I sometimes wondered how often it occurred to cadets that poor behavior toward a girl, for example, or buying merchandise for someone besides a cadet at their store could be ethical issues, issues of right or wrong. A recent graduate confirmed my suspicion when he said, "When someone asked me to buy something for him in the cadet store, it never occurred to me that 'honor' was involved, that cheating was the issue. Ethics and honor are two different things at West Point."

This results in a curious state of affairs indeed. It means that lying, cheating, and stealing become associated with a small number of specific acts and are divorced from a good many others.

On top of all this, some officers actually encourage rule-breaking. When discovered, violations bring swift and certain punishment; but if violations go undetected, rule-breaking is considered by some tactical officers as a mark of initiative. Said one, "I tell cadets, 'I don't mind your breaking the regs, but I'll hang you if you get caught.' I don't tolerate major infractions, but I like to see them use their initiative." One Saturday night after taps three cadets sneaked out of barracks with wet suits and a rubber boat. They made their way to the latticework of steam tunnels that lies beneath the Academy. Emerging near the Officers' Club, they worked their way down to the Hudson and paddled across. Although they landed a mile farther downstream than they had planned, they managed to find their way to Garrison, the small town directly across the Hudson from West Point, where three girls were awaiting them. From there, the group of six proceeded to an all-night party in

Newburgh. The girls delivered the cadets back to the Academy before reveille, and the men attended chapel Sunday morning. "The tac laughed his head off when he heard that," said one cadet. "He thought it was the funniest thing in the world. But if he had caught them, they'd be in confinement to this day."

A more subtle and pernicious form of rule-breaking takes place in the way some tactical officers reward the members of their company for winning a regimental championship in intramural athletics, for example. Though the motivation is perfectly understandable and, indeed, human, the tactical officer who lets his cadets arrange their rooms however they wish or allows seniors to keep their rooms in only a semblance of order is entering into a collusive relationship with cadets. The message is: If you're the best, if you come out on top, you don't have to abide by the same regulations others do. A tactical officer who did this might well be reprimanded by his regimental commander—if the commander found out. No wonder a common cadet phrase is "Keep it in the company."

I suspect some officers encourage rule-breaking because they share the same feeling of powerlessness cadets do. A cadet who violates the regs is only acting out his feelings in a way every officer can understand. A rather dramatic example of this occurred in the spring of 1972, when a cadet broke an informal rule, an old Academy tradition, and refused to buy the class ring. He was called to account by his regimental commander, a colonel. When he still refused, the cadet was ordered, as punishment, to report to his tactical officer, a major, for an hour of "counseling" each day. The major spent the counseling hours playing cards with the senior, thereby supporting the cadet's act of defiance and independence and, at the same time, revealing his own feelings.

But something is curiously wrong when rule-breaking is considered a mark of initiative. If such acts are actually encouraged in an institution in which cadets have a narrow definition of honor, fierce personal loyalties, and a set of restrictions that makes life intolerable, it is not hard to see why lying and cheating are so common at West Point. Cadets quickly discover the truth about honor: in practice it is relative, not absolute, as they had been led to believe.

Honor code violations occur frequently, most West Point officers agree. A 1962 West Point graduate teaching in the mathematics department said, "It took me a long time to realize that the concept of honor has changed since I was here. I'm sure there's more cheating going on. I wish I knew the reasons. One is that kids coming in today aren't impressed with tradition. They're more willing to question and test authority, to violate it. I think, too, that the input has changed. West Point in 1962 had about 2,500 cadets. Now it has 4,000. The overall quality has gone down as the number of cadets has gone up."

A major in the social science department voiced similar feelings: "There's more cheating than when I was here. I'm scrupulous about letting cadets know in class what is and what's not an honor code violation for the work I give them. I cover myself by writing down whatever I've said. I suspect there's going to be a major scandal of some sort soon."

These officers' fears were justified, though I think they were probably naive to imagine there was less cheating during their own West Point days. In April 1973, a major cheating scandal involving twenty-one cadets was uncovered at West Point. All of the implicated cadets, who were charged with cheating in a physics course, were subsequently dismissed from the Academy. It made big news, but it hardly received the attention the cheating scandal of 1951 did when over ninety cadets, most of them football players, were expelled. Cheating is not uncommon at West Point, but only the proper combination of circumstances brings it to public attention.

The coexistence of lying and cheating with a rigid definition of honor is one of the paradoxes of West Point life. So is what cadets regard as the abuse of the honor code by officers themselves. A constant complaint by cadets is that the honor code is not really theirs but is used against them; they feel that officers often use it to enforce the regulations. One example of this is hallowed by tradition and therefore elicits no complaint. Cadets, by regulation, cannot be married. Every time a cadet returns from leave, he must state, on his honor, that he is not married. His honor is thus used to enforce the regulation. A flagrant but unsuccessful attempt to enforce the regulations through the honor code occurred in 1972. Some cadets, chafing at the regulation that decreed that their hair could be no longer than three inches on top, let their hair grow, bought grease to plaster it down, and were thus able to disguise their violation. A tactical officer, angry at this, suggested that a violation of the haircut regulation could be detected by asking a suspected cadet —on his honor—whether or not he was using grease. If he said yes, his hair length would be checked. If he said no, it would also be checked, and if he was found to be lying, he would be dismissed from the Academy on an honor violation. Cadets justifiably thought that such misuse of the honor code made the punishment far more serious than the crime, and they felt insulted that an officer would ask them a question "on their honor," then check them anyway.

Cadets are aware of the contradiction between the principle of trust that underlies the honor code and the frequent mistrust and constant supervision of their behavior. A cadet offered one representative example: "If a waiter in the mess hall reports you to the officer-in-charge, and it comes to the point where it's the waiter's word against yours, the waiter always has it."

Cadets' commitment to the honor code is further undermined by officers themselves who flaunt the code. A major saw two cadets drinking illegally off-post. He knew the name of only one, however, and was unable to report the other. Reassuring the cadet whom he knew that his drinking companion would not be turned in, the officer asked for the companion's name "out of curiosity." The next day the drinking companion was reported for violating the regulations. Incredulous and angry, the cadet who had satisfied the major's curiosity confronted him. The officer reportedly said, "I'm sorry, but things have to work that way sometimes." "It's terrible," said the cadet whose confidence was violated. "I'll never be like that. And I'll never turn in another cadet on honor. Why should I be responsible for throwing a man out of the Academy for something trivial when an officer can lie?"

A tactical officer habitually told his cadet company commander he would not be down to inspect the company on a certain day, then would appear anyway. "It's the kind of thing that would be an honor violation if cadets did it," said one cadet. "People in the company were really bitter about it."

Nor was the example of a man like Brigadier General Samuel Koster, who was relieved as superintendent of West Point in 1970 for his role in the cover-up of My Lai, lost upon cadets. Lucian Truscott IV, a 1969 Academy graduate, wrote ironically of "formerly Major General, currently Brigadier General Samuel Koster, who presided at several thousand feet in a helicopter and at least an arm's length in subsequent months, over the massacre at My Lai. Koster went on to become superintendent of the Academy, where, each and every Sunday, along with the cadets, he recited in chapel the cadet prayer."[2]

Small wonder that many cadets become quickly disillusioned with the honor system. They cannot entirely disregard the code, because violations of it are so serious. But, as a recent graduate put it, "A guy comes here full of ideals. But he soon sees what the reality is and says, 'The heck with it. I'm going to do what I want, and everyone else can do what they want. I won't let myself be caught by an honor code violation, and I'm not going to turn anyone in either, especially a friend.' "

Cadets thus establish their own standards of morality, values that are not always the same as the ideals of the Academy. West Point exerts every conceivable pressure on cadets to mold them into a highly co-hesive, elite body whose foundations are friendship and personal loyalty. It then asks these same men to report each other for honor code violations, an impossible demand. Cadets have a difficult time identify-ing with the ideals their officers exhort, because they perceive the officers as severe, punitive, and sometimes very hypocritical. Finally, the arbitrary manner in which the Academy presents the honor code—*"complete* integrity of word and deed"—to cadets is unrealistic. Ideals are, or can be, ennobling, and this is their proper function. But when they are enforced in a way that leaves little room for the inevitability of failure, they become meaningless. Even the Academy recognizes the impossibility of its task. The concept of "social honor" and, especially, the establishment of two separate codes of behavior—one associated with the honor code, the other with the regulations—show it. The overall lesson that cadets learn is that a lie is a lie is a lie—under certain conditions.

The results of the West Point ethical system are not especially happy. As a 1962 graduate put it, "I think West Point substitutes its concept of honor for morality. And once you start talking about morality and honor as different things, it's easier to separate them a little further in different situations. When you're in a unit in the army, what's consid-ered honor is loyalty to the other people in it. If you're asked to send in a false readiness report, you do it; you do it because you're loyal and because it's for the good of the unit. But by this time no one seems to realize that lying has become honorable. You see the beginnings of this problem here at West Point. The minute one cadet lies for or cheats with a friend, honor becomes contingent on loyalty, on personal whims."

A narrow concept of honor combined with powerful ties of friend-ship inevitably leads to strong internal group standards which are con-cerned exclusively with the interests and self-preservation of the group. Cover-up is one serious consequence of this system.

Cover-up is prevalent for another reason. Cadets are well aware that the slightest breach of duty means punishment, but they find it difficult, if not impossible, to live in an error-free manner. They often find it easier to cover up their mistakes and errors. This is what one former cadet meant when he said, "You had to lie to survive at West Point."

I shall never forget one of the last conversations I had with a West Point officer, a 1962 graduate teaching in the Department of Military Psychology and Leadership. We were walking down a hospital corridor together late one afternoon. He asked me if I was going to a meeting for officers the next day. I told him no and the reason why. He stopped me and said, "That's not good enough. You don't think like an officer yet." He proceeded to give me three different reasons why he himself was not going. Were any of them true? I asked. "Of course not," he replied, "but that's what guys learn here—to cover themselves in every situation. Always have a good excuse; never be caught out."

It is hardly necessary to point out the prevalence of cover-up in the army or where this practice has led us in recent years. I hesitate to say that West Point is responsible for this situation or that every cadet learns it there. But many do, and I have no doubt that the Academy's unworkable code of honor contributes largely to that lesson.

10

WEST POINTERS
AND WOMEN

The manner in which cadets and officers relate to women—dates, girl friends, and wives—adds another dimension to the portrait of a West Pointer. The Academy's martial atmosphere presents a serious obstacle to the easy male-female relationships civilian students often take for granted. "Bringing a date to West Point is like asking a girl to participate in a weekend-long inspection," complained a junior. A recent graduate recalled, "If you really liked a girl, you wouldn't let her visit you at West Point; instead, you'd meet her in New York or in Newburgh. I really didn't enjoy having a girl up because of the atmosphere. It just put a damper on everything. You couldn't reach over and touch her in the movie without feeling that an officer was behind you with a 2–1 pad [punishment] in his hand."

Cadets are hard put to find privacy except at Target Hill Field, which —as one cadet observed—"is like a drive-in without a screen." In fall, spring, and summer Flirtation Walk, a mile-long path along the Hudson which is off-limits to officers, is also a favorite spot for privacy. As one junior said, "On good days in the spring, the action is so heavy in the bushes on 'Flirty' that you don't dare step off the path." "Flirty," as cadets call it, and Target Hill Field are the unofficially recognized locations on post where boys will be boys. In public, cadets may not so much as take a woman's arm without risking punishment, but feelings are much more unrestrained in places well out of the public's view. West Point is a remnant of the Victorian world, where appearance and decorum force passion and feeling underground.*

Paradoxically, the Academy fosters both chauvinism toward and dependence on women. Both traits are well established by the time a cadet becomes an officer and a husband.

The interaction between cadets and women often begins in the office of the cadet hostess at the Point. Occupied by several women who are widows of West Pointers, this suite of offices is a place cadets wander into during the day for coffee, doughnuts, and informal chat with other cadets or the hostesses themselves. This is also the place to which letters, and often photographs, from girls who want blind dates with West Point cadets are channeled. The hostesses transcribe the vital information from these letters—the girl's name, address, age, height, color of hair—onto a 3 x 5 file card which is then kept on record in the office. If a cadet wants a date, he simply goes to the file, riffles through it, makes his selection, and either writes or phones the girl. Certain cadets (called "Hostess Twinkies" by their peers) who become well-known to the cadet hostesses receive special favors from them, such as the opportunity to attend debutante balls in New York or to escort the daughters of visiting dignitaries to West Point.

For large dances—called "hops" at West Point—a great number of blind dates are required for the cadets. The cadet hostesses make arrangements to bring girls from surrounding colleges—Marymount, Manhattanville, or Ladycliff. They are transported to West Point on a large bus. As one cadet said, "There's something dehumanizing about it. Cadets drift in and out as they please. There's no sense of responsibility toward the girls. 'They came on a bus, they can go back on a bus' is the prevailing attitude."

Although busing at West Point may be dehumanizing, the ultimate expression of insensitivity toward women occurs during the first class trip, when fifty to a hundred cadets travel to another army post such as Fort Bragg or Fort Knox. The post commander usually holds a party and reception for the cadets in the evening. Again, blind dates are required for cadets. Before the reception, cadets place a dollar each in

* When sexuality is institutionally suppressed, one wonders where it surfaces. Perhaps during inspection in ranks, where cadets are ordered to "pop it up" (which means "expand your chest"), "make a dirty movement" ("straighten your back"), or "ram it in" ("pull your chin in toward your chest").

103

what they call the "pig pool." The cadet with the ugliest date collects the pot. The judge is an officer from West Point who accompanies the cadets on the trip. "If you think you have a chance of winning the pig pool," said a cadet, "you take your date and introduce her to the officer. He takes note and decides after the party which cadet has the ugliest date. Sometimes there's a second and third prize, too. Guys can make seventy to eighty dollars at times."

As at all men's colleges, women are the favorite topic of dormitory (or barracks) bull sessions, where all the bravado of young men experienced in fantasy and immature in reality is displayed. The conquest, or fabricated conquest, of a woman is of great moment for cadets gathered together after closing hours. During Beast Barracks, upperclassmen have been known to rate new cadets on a "stud index," a ratio expressing the number of women a man has had relations with over his chronological age. The ideal is evidently 1 : 1. Cadets express astonishment, however, when women talk about them the way they talk about women. A cadet was standing behind two girls watching a parade one afternoon at West Point when he saw one turn to the other and say, "Jesus, Sue, will you look at that! There must be two miles of meat marching out there." The cadet sought out an officer he knew, a married lieutenant colonel, and asked, "Sir, do women really talk that way?"

Every cadet yearns for the weekend, which brings the opportunity to live just a little and, perhaps, to date. Whether his weekend release is conducive to healthy man-woman relationships is highly debatable. The serious part of a cadet's life occupies the entire week, Monday through Saturday morning. Weekends are playtime and girls part of the sport. Women come after the mission is completed. I am not saying that women are not important to cadets or that cadets are not serious about them. Of course they can be. What I believe is that cadets internalize the Academy's priorities in ways that show up clearly after they become officers and husbands.

It is also true that the rigid, punitive, military atmosphere of West Point makes it difficult for cadets to relax, to be themselves, with women. I suspect the result of this is that cadets and girl friends come to know each other much less well than do men and women on civilian campuses. And perhaps, knowing them less well, cadets have a harder time recognizing women as people. An officer's wife observed, "Cadets read *The Invisible Man* in their English classes, but I don't think they know they treat women as invisible. When I meet cadets, they usually say 'Ma'am' and look at my eyebrows."

No wonder that one senior, a bright, sophisticated individual only a month short of graduation, said, "I'm able to handle myself pretty well around other guys, but I still don't feel comfortable around women yet."

The regulations at West Point, which prohibit cadets from owning (or for two years from even driving) cars, leaving post on many weekends, or buying liquor, cause them to be more dependent on their dates than are students at civilian colleges. A 1972 graduate said, "Girls usually came to the Academy, especially during our first and second year, instead of our going to them. If you wanted to use a car, they had to bring it. If you wanted liquor, they had to buy it. When I date now, women tell me I expect more from them than other guys [civilians] do." Cadets must be back to their barracks by 12:50 A.M. on Friday and Saturday nights. *They* have the closing hours, not the girls. It was surprising to see cadets—the models of masculinity during the week— kissing their girls goodbye and getting out of the car, after which the girls drove off. "It was kind of like the tables were turned," said a cadet. "Like we switched roles. I hated it and vowed it would never happen again."

I was rather surprised by the amount of insecurity among cadets about their future dating opportunities. They feel that West Point is their last chance to date college girls. Fort Benning, Fort Bliss, and Fort Polk are, they know, minimally coeducational. Moreover, constant companionship is a big fact of life at West Point. Cadets are always together and have little chance to be alone. The prospect of living alone, suddenly, at some isolated post after graduation is frightening for many of them. As a senior said, "The idea of being alone somewhere is depressing. A lot of guys get married to avoid it."

A cadet's desire to get married is sometimes carried to desperate lengths. One senior, planning to be married the day after graduation in June 1972, broke up with his girl friend in April. He kept his reservation in the chapel open, however, in hopes of finding someone to replace her in time. "Cadets think they'll lose a security blanket unless they find somebody by the time they leave the Academy," said a 1972 graduate. "The kind of experience we have at West Point doesn't prepare us for life on the outside, even in the army." A West Pointer who went to medical school after graduating from the Academy said, "I didn't get married for my first year out because I wanted to prove to myself that I could make it on the outside. I wondered if I'd know what to do if every minute weren't planned. I'd never been on my own before; I had lived with my parents until I went to West Point, and you can't call being at the Academy being on your own."

Speaking of his classmates' haste to marry, one cadet observed,

"Dating is an important escape here. The atmosphere is so strained that cadets are desperate. They dupe themselves into thinking they're in love. Cadets are like college girls who worry they'll end up as old maids unless they get married by the time they leave school."

The Academy's atmosphere also seemed to create relationships between men and women which were strongly obligatory. Even if his feelings were not quite so strong toward her, could a cadet easily dump a girl who had loyally come to see him, with a car, for two or three years? I was impressed how often the word "devoted" came up when a cadet talked resignedly of his girl friend and their impending marriage in June.

A quarter of each year's graduating class rushes into marriage within two months of graduation. The chapel is booked up for months in advance. For three days immediately following graduation, weddings are held every half hour throughout the day. "It's *very* impersonal," observed a cadet. Fifty percent of the class is married within a year of graduation.

In the middle of my own tour at West Point, I gave a talk on sexuality and marriage to a group of seniors and their fiancées who planned to be married in the Catholic chapel immediately after graduation. As I looked around the large, crowded room and listened to the questions that were asked me, my doubts about the readiness of these cadets for marriage grew. I remembered what a lieutenant colonel's wife had said about them: "Cadets should be required to wait three years before they marry. They can't really handle it. I saw them at Fort Benning right after graduation when we were stationed there several years ago. When they get out of here they're just like freshmen who're away from home for the first time. They're not ready for lots of things and certainly not for marriage."

The Academy, though often treating its cadets like children, insists they act like adults from the moment they enter West Point. I once asked an officer what West Point training gives an individual. "Varnish," he said. The result is often a product that outwardly is mature and poised, but, behind the facade, is often insecure and dependent. Stress sometimes revealed what lay underneath.

One of the saddest stories I heard at West Point involved a cadet and his girl friend, the daughter of a retired noncommissioned army officer. They had dated for nine months. A week before graduation the cadet told the girl he would come back to New York after the summer to get her, that he could not leave her behind. That night, for the first time (in her words), "I didn't stop him. I don't know why. I guess I wanted to show him how I felt. Two weeks later he said goodbye and told me I didn't talk enough for him—I never had—and that he couldn't stand to spend the rest of his life with someone like me who couldn't 'open up.' The nine months we went together he had never mentioned that."

Unfortunately, the girl found out she was pregnant several weeks later. She wrote the cadet, by now a second lieutenant. He called her back immediately after he got her letter. At first he denied that the child was his, then changed his mind and said that he would send money for an abortion. The girl refused to have one. He offered to marry her and said he would fly up in a couple of days. The girl told him to sleep on it and talk it over with his parents.

The next day the officer's mother called the girl and said she had grave doubts that her son was the father of the baby. As the girl said, "She gave me the impression that John was normal and healthy to try for nine months, but that I was a tramp to give into someone I loved and trusted. After I talked to her for a few minutes, I asked to talk to John. He was a totally different person. He told me how stupid I was to go through with the pregnancy and said I'd get no help from him. He said I was not to mention his name at all; I'm supposed to tell my parents and the doctor that I don't know who the father is. He told me I'd never hear from him again and hung up. Actually, he gave the phone back to his mother. She tried to talk me into having an abortion. Finally she said John would send me some money, only to protect his honor. Not because he cared for me or was admitting anything, but only because he'd gone to bed with me."

Of course incidents like this happen every day, everywhere. But it jars one slightly to hear that a second lieutenant fresh out of West Point would so quickly run to his mother in a time of stress.

The Academy also rather inadvertently encourages a curious kind of irresponsibility toward women. Whenever cadets return to post after leave, for example, they must state on their honor that they are not married. One cannot be married and be a cadet. But a cadet has never been asked to resign from the Academy for fathering a child. Cadets receive a straightforward message: preserve appearances, no matter what, under pain of expulsion. It is almost a joke around West Point that every June several women with small babies watch their future husbands graduate.

I suppose, too, that women represent a threat to the exclusively male military world. There is an unspoken fear that a woman will divide a man's loyalty to the mission and to the institution. In an act which has enormous symbolic meaning, the Academy holds a Ring Ceremony each May for juniors. Solemnly, each cadet receives a West Point class ring—

signifying marriage to the institution—which he usually wears on the third finger of his left hand. If the graduate later purchases a wedding band, he has it bent to conform to the shape of the West Point ring and wears both rings on the same finger. His wife often wears a miniature class ring and a wedding band in the same fashion, indicating her own devotion to both West Point and her husband.

West Pointers, products of rigid training, often made rigid, controlled husbands, men whose preoccupation with image often overrode any other consideration. "I wouldn't be seen in uniform carrying a child, holding an umbrella, or carrying groceries," said a major. "It doesn't look good for an officer." Wives whose husbands adhered to this informal but strong tradition would occasionally be seen struggling out of the West Point commissary with the combined burden of both children and groceries while their husbands' hands were completely free.

This desire to maintain appearances at all costs sometimes extended to an unwillingness to admit problems even within their own family. "It *is* difficult," one major said. "You can't confess to any problems at your job or in your family. If you're having trouble at home, it means you're inadequate, and it may show up on your efficiency report." A West Point psychologist observed, "The number one problem on post is marital difficulty, and officers will never come for help, because they're afraid to expose their failures."

Officers rarely sought psychiatric help on their own. Like cadets, they worried about confidentiality. This was unfortunate, since some wanted help, such as the 1962 West Point graduate who confided to me at a party, "I'm in a tight spot—at work and at home—and I'd like someone to talk to, like a psychiatrist. Right now I'm ready to blow up. But if the army finds out you've seen a psychiatrist, it means you're not suitable officer material." More alarming was what one wife told an Academy psychiatrist: "My husband is thinking about suicide, but he doesn't want to come for help because he's afraid it will go on his record."

Although officers were reluctant to turn to a military psychiatrist because they thought it might hurt their careers, their wives were less concerned about such matters and came for help with problems that women consult psychiatrists for everywhere: anxiety, depression, psychosomatic symptoms, and, especially, trouble with their husbands or children.

The women who came to the psychiatry clinic because of their marital problems generally portrayed their husbands as insensitive men who treated them more as a member of a detail than as a wife and partner. A cadet recalled hearing a tactical officer communicate with his wife one afternoon: "Wife, this is Duncan. I'm messing with the troops. Home at

2100 hours. Out." One of the most distraught women I saw at West Point complained that she felt like a slave, that her husband expected her to do every bit of work around the house, raise the children, and serve him a drink when he came home after work. The officer himself was basically a very dependent person who told his wife he would shoot himself if she ever left him. She felt understandably trapped.

Wives reported that officers generally had trouble expressing feelings. Indeed, many officers thought that feelings, especially tenderness and sadness, were signs of effeteness. Naturally, this attitude sometimes led to marital problems. It seemed as if a West Pointer's wife ended up feeling for both her husband and herself. There were many symptomatic women but few symptomatic officers. When a West Pointer was confronted with his wife's emotional problem (in reality their marital difficulty), he had difficulty accepting his wife's feelings. "Officers think wives are putting on when they really have problems," said a doctor at the Academy. "They think, 'She could pull herself together if she'd only try harder.' "

One problem for women was the emphasis West Pointers placed on masculinity. "There's a big machismo thing among West Pointers," commented one officer's wife. "They see themselves as big, virile, aggressive men who won't do anything that looks like women's work." An obstetrician assigned to the army hospital at West Point was convinced that there was more sexual activity among West Point officers than among any other comparable civilian group of men he had known. Another doctor added, "I think lots of officers here have sex compulsively every morning because they feel it's healthy. It's good to keep the tubes clear. It's a ritual, like calisthenics."

One West Pointer's wife complained to her doctor that her husband had refused to have intercourse with her since he had injured his shoulder. He had a large cast on his shoulder and upper chest which made it impossible for him to assume his customary position during love-making, and he felt it was unmanly for him to have intercourse while lying on his back.

Sometimes the "masculinity" shaded into authoritarianism. One lieutenant colonel's wife told her husband she had overdrawn their mutual checking account by five dollars. At the time her husband said nothing, but several days later she found out he'd gone to the bank and tried to have her name removed from their joint account. "That was his way of giving me an article 15,"* she said.

---

* An official reprimand given to officers, NCO's, and enlisted personnel for misbehavior.

While I saw many good fathers at West Point, I was rather surprised by the formality between officers and their children. They often insisted that their children call all adults, including themselves, "sir" or "ma'am," and often addressed their own sons as "sir." We did not see many officers' children in the psychiatry clinic. From the reports of some wives, however, I formed the impression that an officer's desire for his son to be competitive was sometimes a major source of tension between him and his son. "There's a big jock thing here at West Point," said a major's wife. "When West Pointers get together, they're always jocks with each other. There's lots of emphasis on sports and scouts. They push their sons, who often wind up as competitive as their fathers. But it's hard for a lot of kids who aren't naturally athletic to have to live under that kind of pressure." Women also reported that relationships between fathers and sons were strained when a husband returned from a tour overseas. A boy who cried, who ran to his mother, might be considered a sissy, a mama's boy. Some officers would not tolerate it and treated their sons much more severely than they had before.

A continual debate was carried on among members of the psychiatry service at West Point over the question: "Is the incidence of mental health problems higher in military or civilian families?" One psychiatrist was convinced that the military environment was pathogenic: officers were overcontrolled and insensitive, he insisted, and cited the West Point graduate who went off on a hunting trip the morning his wife had a breast biopsy. The psychiatrist, incredulous when he heard the officer's plans, asked him if he wasn't concerned about his wife. "Sure," replied the major, "but why should I worry? If she turns out to have cancer there's nothing I can do, and if she doesn't there's nothing to worry about."

Divorce within army families was extremely rare at West Point during 1970–72. It was my impression that West Point marriages are in fact more durable than civilian marriages—for several reasons. Officers know that a wife is usually an asset to their career. Some positions—an assignment to West Point as a tactical officer, for example—are not even available to single men. Most army posts are family-oriented; an unmarried man is often forced to lead the solitary, lonely life of an outsider. Also, a sense of responsibility and duty toward dependents is strong among West Pointers, and divorce is considered a breach of duty. West Pointers are, furthermore, reluctant to expose their failures; many feel that a marital breakup would be bad for their career. But though the incidence of divorce is low, there is no doubt that an officer's West Point background, as well as the environment of West Point itself, cause unique strains for officers' wives and their families.

One unhappy wife came for psychiatric help in the fall of 1971. Her husband's previous assignment had been Vietnam. Their relationship had deteriorated since they had come to the Academy the year before. She saw her husband infrequently. He spent long hours at work and, in his free time, was involved in an endless round of squash in winter, tennis in summer. Also, her husband had become stricter with her two young boys in the last year. "When he's around home at meals, I feel like we're in cadet mess. I think he expects the kids to behave like cadets; they have to sit up straight and not talk. If they spill any food or milk, they have to leave the table."

Women often had a difficult time at West Point, largely because of the military world itself, which gave them even less status than the civilian world did. "If the army had wanted us to have a wife, they would have issued us one," said a major in the mathematics department. An officer's wife was more direct: "Words like 'stud' and 'pussy' show exactly what the military attitude toward women is." Women in situations independent of their husbands or boy friends are perceived as threatening, perhaps seductive, by military men. A thirty-year-old major's wife was doing research in the stacks on the second floor of the West Point library. She was requested to leave with the explanation, "In order to protect cadets from aggressive females, there's a rule that women are not allowed to loiter on the second floor." After the woman protested, one of the head librarians called the next day and apologized. The rule, he said, was really made to protect women from aggressive cadets. Recently, he added, a colonel's daughter had been attacked in the library by a "man in gray." Another woman, a colonel's wife, who visited the stacks, was told she should remain standing; sitting was apparently considered too provocative. At an Academy party, a colonel came up to a major's wife and, in a genuine attempt to compliment her, said, "My, you're a pretty decoration tonight."

Women and children share the same designation in the military: dependents. The consequences are sometimes humiliating. One wife who took her small child to the West Point Child Care Center and signed her name in the space marked "parent" was informed that that meant her husband's name. Another wife recalled, "A couple of weeks ago my identity card expired. I couldn't go down by myself and renew it. My husband had to come with me and sign a form saying I was his wife."

These incidents make West Point wives feel, as one said, "like a nonperson." Her own identity is secondary to her husband's career. She finds herself in the traditional roles of homemaker, child-rearer, and social facilitator. Some are understandably disillusioned. "You learn at an early age you're going to take care of a family and live with a husband," said one, "and suddenly you realize the family, the kids, and all the responsibility are yours—and that's *all*. You end up hating the army more than you do the husband." Some determined women were, of course, able to fashion a more satisfactory life for themselves that did not fit the traditional stereotype, but it was not always easy. Army life means frequent moves, and army wives are not always encouraged to make careers. "West Point is better than most places that way," said one woman. "At Fort Sill, for example, a wife who works is looked down upon."

But even for those who are able to establish a life for themselves, there are apt to be conflicts. The wife of a lieutenant colonel at West Point who has intermittently pursued her career while moving from post to post commented, "I don't know if there's much of a role for an independent wife in the army. There are times when I feel I've let my husband down by being different and going my own way. He's assured me it's no problem, but that's certainly not the typical reaction of most husbands here. I'm lucky. At West Point there's more talk about a different role for women than there is evidence of a role itself, which has always seemed to be traditional and oriented toward group and social activities." This woman, as she knew, was an exception. I have little doubt that many officers' wives, for the most part bright and educated, feel restricted and frustrated within their role.

The rank structure of the army also keeps some women boxed in. Competitive with wives of their husband's rank, unable to share their thoughts and feelings with women whose husbands were of lower rank, and afraid of exposing themselves to wives of their husband's superiors, a couple of them confided to me that the only person they could talk with freely was their cleaning lady.

When a West Point officer goes overseas for a year or when one looks closely at what goes on within his family, however, the irony of a woman's status as a "dependent" within the military becomes clear. While her husband is away, she assumes total responsibility for the family. In fact, the most serious conflicts in a West Pointer's marriage often surface after he returns. His wife, more willing to assert her independence after a year of responsibility, sometimes has trouble resuming her traditional role. A woman who came to West Point with her family and husband immediately after he had completed a tour of duty in Vietnam in 1970 said, "We had lots of trouble at first. He kept saying, 'You're

112

so damned independent.' To be perfectly honest, I wanted him to extend for six months. I realized I'd have to get back into the routine. I had my own freedom and wasn't indebted to him for giving it to me. I began thinking and came to the conclusion an overseas tour isn't that much of a sweat. Children don't see that much of their daddies anyway in the army; the women really bring up the kids."

Not all women, by any means, wished their husbands would extend their tour of duty. Most couples hated to break up and were only too glad (and relieved) to see each other again after a year apart. Yet even in those families, serious readjustment problems occurred, especially if the wife insisted on demonstrating to her husband or outsiders her competence, independence, and obvious ability to care for the entire family.

It has been said that the army is a career for men who wish to remain boys. While genuinely mature military men do exist, of course, it was often difficult not to view the West Point officer as an incompletely developed individual, one who had found in the institutions of West Point and marriage surrogate parents to replace those he had left behind in the civilian world. As a cadet, he found in the Academy a strict, demanding father, one quick to punish but one also able to reward him. In their relationships with their West Point husbands, officers' wives sometimes reminded me of indulgent mothers who granted their sons responsibility but were really quite capable of handling matters on their own. The West Pointer achieves security this way, but at a high cost to his independence and maturity.

11

CHANGES

In seeking to mold a different kind of man, a West Point officer, the Academy has been compelled to create another world in miniature. The walls of the Academy are high, the facade imposing, but into this sanctuary the world will inevitably enter. Efforts to create a world divorced from civilian reality are never entirely successful.

Nor, in fairness to the Academy, does it exactly wish to be. A totally hermetic world would undermine its mission: to develop an officer who is prepared to lead the smallest combat unit or to advise the highest governmental council. (The pretensions of West Point's British counterpart, Sandhurst, are considerably more modest. "The aim of Sandhurst today is the same as twenty or thirty years ago—to produce a chap capable of commanding a platoon of soldiers," said a British lieutenant colonel teaching there.[1]) Isolated military training would never prepare future graduates for the multiple roles the Academy envisages they will later play within the army and, indeed, within government. West Point requires educational interchange with the civilian world. Virtually all West Point instructors boast of civilian academic credentials in addition to their West Point degrees. The Academy explicitly prepares cadets for future graduate study in civilian universities, and seventy-five percent of

114

Academy graduates go on to graduate school sooner or later in their careers.[2]

In the last ten years, West Point has altered its academic program greatly. Cadets may now take elective courses, of which there are well over a hundred. They may concentrate, though not major, in certain areas of work. They have more opportunity to do honors work and independent study or research projects. The academic departments at West Point are themselves becoming more specialized. While art, music, and philosophy are still tucked away in the English department ("English takes care of philosophy here," said a math instructor), chemistry is now a separate department from physics, as is history from the other social sciences.

Changes in the curriculum over the last decade have been paralleled to a lesser extent by changes in the training program. Cadets are granted more free weekend leaves and the right to wear civilian clothes at certain times and places. Plebes are harassed less than they were in the past, and all cadets have fewer parades each week.

Above and beyond these changes, however, the Academy is faced with the great and current issues of today: demands for more individual freedom; an insistence that the constitutional rights of blacks and women as well as the legal rights of all individuals, including cadets, be protected. Where these issues seem to threaten its special domain of expertise—military training—the Academy has not always been a gracious host.

In December 1972, compulsory chapel for cadets was ended after the Supreme Court ruled that mandatory religious services at all the military academies were unconstitutional. Until then, cadets were required to attend one service or another. If, for example, a Jew wished to attend the Protestant service, he had to make an appearance at his own service first. Cadets who claimed to have no religious affiliation were forced to attend ethics classes on Sunday morning. After mandatory chapel was eliminated, cadets were certain—and wrong, as it turned out—that West Point would reschedule the Sunday morning time for some other activity. "We now have Sunday mornings free," said a senior. "It's great. We were sure the Academy would substitute religious classes in place of the services, but it hasn't."

Since 1969, thirty-five to forty blacks have been admitted to the Academy each year, a marked increase over the five to ten admitted each year in previous classes.[3] Blacks come to West Point not only for many of the same reasons whites do—education, prestige, security, status—but also for reasons of their own. I had the impression that blacks, much more than whites, perceive West Point as a vast step upward on the social

115

scale. It not only means improved social standing, but also the opportunity, in many cases, to escape from a chaotic social and family background into a world of order and discipline. Also, some come with a wish to transcend their blackness. The Academy, they hope, will give them freedom, equality, and a firm identity they would not achieve on the outside. For these blacks, I think, the Academy is sometimes a disappointment. Blacks already at the Academy will not let a new black escape his identity or exchange it for a new one. "In fact," said one, "I was pretty neutral when I came here. I really learned about myself and blackness after I got to the Academy."

Black upperclassmen do not approach new black cadets when they first arrive at West Point. "We figure," said a black cadet, "that we can let 'em suck a little bit and see what it's really like; we'll let 'em go through Beast Barracks and get some pride. They'll join the frat soon enough anyway." A black's socialization into West Point's black culture begins after reorganization week and proceeds apace with his military indoctrination.

New blacks get the message that blacks should stick together at West Point. They are told about discrimination at the Academy: how West Point takes no pains to bring black girls to West Point for black cadets as it does white girls for white cadets. (The fact is that few black girls attend the colleges from which West Point recruits dates for cadets. The Academy has made a gesture of responsibility toward blacks in the social area by employing a black cadet hostess.) Blacks also cite what they consider an unofficial policy on the part of white officers which discourages black cadets from dating black enlisted women stationed at West Point. Cadets should consider themselves officers, they are told, and not date enlisted women. "New blacks are also told," said a white officer, "that to survive here, they have to associate with other blacks." This may be true in some cases. "What bothers us," said a black cadet, "is that when a black plebe is D, other blacks don't know about it until it's too late. If we knew earlier, we could do more to help."

Discrimination against blacks has always been prominent in West Point history. Traditionally, blacks were isolated, burdened with extra restrictions, subjected to verbal abuse, and given few opportunities to participate in sports or social activities.[4] In recent years, specific incidents have become rare, though they still occur. One black officer told of several white cadets running through the halls with sheets over their heads, dressed like Klansmen, trying to scare blacks. And at Camp Buckner, during the summer of 1971, a black sergeant was portrayed as an Amos 'n' Andy figure.

"But the real problem," this officer continued, "is that black cadets feel they're losing their identity. They feel they don't belong at West Point, that they're outsiders at a white institution, that they're marginal men. They don't see West Point, the army, as theirs. One said to me, when we were talking about Vietnam, 'Why should we have to fight wars for whites?' "

One black cadet, an outstanding athlete who resigned in 1972 after his second year, confirmed that the commonest complaint among blacks at West Point was "loss of identity" rather than discrimination. "At West Point you have the constant feeling you'll lose what you want to keep. The way of relating to friends back home is an example. After a while you don't even speak the same language as they do. My friends back in Jersey don't even know what I'm talking about sometimes. The longer you stay here, the more isolated you feel; that's why we blacks stick together. I know the other guys [whites] get nervous when they see us all together, but what do they have to complain about? Nobody objects when they sit together. But you still feel out of touch with your own people. A friend of mine in high school who's now a Black Panther said to me when I came home on Christmas leave last year, 'I hope you and I are never on opposite sides.' That kind of talk makes it hard to stay here."

He continued, "I haven't had any bad scenes with white guys since I've been here. The crap, the regulations, apply to everyone. But it's hard socially. I've dated white girls a couple of times, but you really don't like to do it. You realize that you're going to make some girl pretty happy after you're an officer, and if you marry a white that means you're taking it away from a black girl. But there aren't that many black girls around, and the ones who are get old—the circuit riders. It's hard to date blacks at someplace like Vassar, because there are plenty of black guys at colleges that have black girls, so why would girls want to come here? Besides, when they hear of a black guy at West Point, they think there must be something wrong with him."

The most telling account of the black-white issue at West Point came from a white officer who had taught black history at the Academy. "In 1968," he said, "the Academy decided to dramatically increase the number of minority cadets, which meant blacks. Until then, there were only about twenty black cadets in the whole corps; there are now about forty each year, and they make up about four percent of the corps. The Academy is under pressure to make it ten percent, since West Point is federally supported. This means that we'll actually have to recruit blacks sooner or later. At present, the Academy only encourages blacks to come after they've applied. This sudden appearance of blacks in an almost totally white culture has led to a number of problems that the Academy has never had to face before. To what extent is black hair different from white hair? Should the haircut policy be changed? Should changes be made in the curriculum? What kinds of changes? To what extent should blacks be kept together or dispersed within the corps?

"The Academy has operated from the best of the late 'fifties' motivation: if you put people together, racism will disappear. Blacks don't share that view. They feel that West Point has to make changes. That's not unreasonable. The blacks here haven't been radical or militant, but they're asking for reforms that would allow them to express their uniqueness. Naturally, whites don't like it. The haircut regs have been changed so blacks can have Afros, which means their hair can be longer than whites'. Whites don't like all-black parties; they feel this discriminates against them. Tactical officers are under orders to guarantee equality of treatment, and they're so afraid of being accused of racism that they sometimes discriminate in reverse.

"Blacks are a new thing at West Point, and integration has been hard," he went on. "It hasn't been made any easier by the general atmosphere of paranoia at the Academy. White cadets feel that everyone is against them anyway. They see blacks together and suspect a conspiracy.

"Blacks wanted to have knicknacks, like a Black Power flag, on their walls. When they were told to take them down, they asked, 'Why can whites have confederate flags on their walls?' It was then decided no one could have anything on the walls. Then blacks took to wearing rawhide bands on their wrists. When told to remove them because they detracted from the uniformity of the corps, they asked why white cadets were able to wear prisoner-of-war bracelets. The commandant had to rule arbitrarily that the wearing of POW bracelets was a much different thing from the wearing of rawhide straps.

"The contemporary affairs discussion group is really a black student alliance here," the officer observed. "It's the most organized group around. And because it's organized, it has power. Last year, shortly after his visit to West Point [in 1971], President Nixon proposed that a monument to the Confederacy be erected at the Academy [a monument to the northern dead in the Civil War already exists at West Point]. The blacks, after they found out about this, took very active steps to stop it. They refused to be co-opted. As the matter came to a head, they drafted a manifesto like the Declaration of Independence and listed their problems and grievances. It was signed by every black in the corps and by every black officer except one. When it became obvious the superintendent was

118

not responding, the black cadets threatened mass resignation. That was the point at which General Knowlton encouraged tactical officers, through the commandant, to be more sensitive to blacks.

"That was the first part of the year. The last part centered around organizing the sickle cell fund benefit." This benefit, held in the Academy football stadium in late May 1972, brought several black celebrities to West Point. Proceeds went to the sickle cell fund. A white officer said of the event, "It's the first socially conscious thing that's ever been done at the Academy."

The officer who had taught black history commented on the tensions unique to black cadets at the Academy: "There are difficulties in all directions. Blacks don't feel they can communicate with white cadets because their problems are different. Nor do black cadets feel easy about talking to black officers, whom they perceive as having sold out to the system or possibly jeopardizing them because their allegiance is to the military rather than to other blacks. Black cadets are asking hard questions of themselves: Is it possible to be black and be a cadet? If I'm a cadet, am I abrogating responsibility to my race?"

As for the attitude of whites toward blacks, he remarked, "Frankly, the corps of cadets is more racist than the officers are. Most white cadets, I think, just don't like 'niggers,' and there's nothing in the environment that will cause them to change. The culture of the Academy is basically southern, and there's continuing prejudice, though it's not so obvious as it once was.

"But West Point," he concluded, "is now caught in a situation it can do little about. Now that certain policies have been altered for blacks—haircuts, parties, the fund benefit—there's no turning back. Escalation is built in. Racial problems are in a nonviolent phase right now at the Academy, but I think the situation will probably get worse. Blacks in the future will be more militant, and people in positions of responsibility here at the Academy aren't intelligent, sensitive, or informed enough to deal with them."

Another officer confirmed this. "Black identity is becoming more and more an issue as time goes by. They want more than equality with whites. They want superiority with respect to cohesiveness within their group. They insist that their loyalty to each other transcends loyalty to their squad, platoon, or company. Even the occasional black who wants to dissociate himself can't do it. I think this has led to a certain splintering within the corps. Blacks have challenged the rules and regulations. Upperclass blacks often talk to plebes, and black plebes visit each other in their rooms in the evenings. One guy was even thrown out for per-

119

sistently visiting other blacks' rooms. No one's wanted to confront their challenge for fear of being called a racist, and I think blacks have pretty much got away with what they wanted. I think a showdown is coming. A subgroup like this won't be tolerated at West Point."

A black cadet, however, qualified these views: "Blacks *do* have strong loyalties to each other because of our blackness and our common background. And as more and more blacks have come here, there's been more and more allowance and toleration for us. But we also have strong loyalties to our companies. And since Percy Squire—who provided us with a focus for black loyalty—graduated in 1972, our loyalties to each other have become more dispersed. There's also more allowance now for blacks to express different views and less splintering of the corps than there was in the past."

Regardless of their dissatisfactions, there was no evidence, as of July 1972, that black cadets resigned from the Academy more frequently than whites. A black senior said, "Blacks tend to feel they *have* to stay. There's lots of pressure, especially from parents. You also feel a lot of responsibility to other blacks to make it in the army, one of the great white institutions. When I finish here, I can lend my understanding to black-white problems and be someone blacks can look up to."

However, some unexpected psychiatric statistics on blacks at the Academy became available for the period from July 1972 to September 1973. In that period, three black cadets suffered schizophrenic breakdowns. While schizophrenia is not rare among college students—approximately one in one thousand per year occurs—it is unusual at West Point. In the entire two years I served at the Academy, only two cadets—both white —were diagnosed as schizophrenic. Further, in the summer of 1973, nine out of twenty-three cadets who were diagnosed as having an acute stress reaction—usually incapacitating anxiety or depression—were black. Considering that only about forty blacks entered in the class of 1977, this number was greatly out of proportion: about one-quarter of entering blacks. If a comparable number of whites had had adjustment reactions, it would have amounted to a staggering 235.

One black senior who had a serious but temporary break with reality illustrates the problems which can beset a black who wishes to integrate himself into West Point and deemphasize his blackness. As a West Point doctor recalled, "In 1972, a black cadet graduated who had been a particularly charismatic leader of the other blacks at West Point. He had been able to get support from other blacks because he refused to compromise with the system. He wouldn't let his identity as a black or the identity of blacks as a group at West Point be destroyed. He was one of the people most active in the sickle cell fund drive. After he left, the leadership of the blacks fell upon another man whose background was very different from his predecessor's. This man came from an upper-middle-class family in the South. He'd gone to a Catholic high school where there were only a few other blacks. Most of his friends were white; I don't think he even thought of himself as black. But because he was an exceptional student and was looked up to by the powers-that-be, he was put in a position of leadership.

"He had trouble from the beginning," said the officer. "The blacks expected him to represent them to the administration in the same way his predecessor had. He said, 'I don't really know how to relate to other blacks. Lots of these guys have eaten grits and slept with rats. I've never tasted grits and never seen a rat.' He just wanted to be a cadet and have a career in the army. At any rate, these conflicting pressures finally got to him. A militant Black Muslim leader came up to the Academy and suggested—there were debates afterward about just what he did say—that it was impossible for blacks to serve in the army of the white oppressor. This cadet took it literally and began brooding about it. The incident apparently precipitated his breakdown. At first he was totally mute and couldn't or wouldn't speak to anyone. Then, for a short time, he believed someone was controlling his mind. Eventually he went back to classes and was allowed to graduate because he'd had such a good record up until then, but he didn't receive a commission."

I do not know the cause of the high number of psychiatric casualties among blacks, but I have no doubt that the transition from the black civilian world to West Point is enormously more difficult than the transition from the white civilian world to the Academy.

One new black cadet, not a psychiatric casualty, resigned from the Academy in 1973, though with his sense of humor intact, if the following story is true. Apparently the strain of Beast Barracks finally became too much for him. "One day," an officer remembered, "he started singing in ranks—songs like 'Ole Man River.' An upperclassman pulled him out and started yelling at him. Just then Mrs. Knowlton, the superintendent's wife, walked by. The upperclassman stepped back, embarrassed that he'd been seen yelling at the guy. Mrs. Knowlton asked the black cadet, 'What do you want?' He rolled his eyes back and said, 'I'd like a piece of watermelon.' By that time he didn't care what happened anymore. They let him go two days later."

Statistics bear out the assertion that blacks and whites who enter the Academy together are far apart in attitude. For example, thirty-nine percent of the blacks who entered in 1970 labeled themselves as politically liberal, while only twenty-three percent of whites did. Ninety-two percent of blacks felt that compensatory education should be provided to disadvantaged people, while only fifty-one percent of whites did. Ninety-four percent of blacks felt that more governmental programs to reduce poverty were needed, while only sixty-eight percent of white cadets did. Ninety-four percent of blacks supported school desegregation; only forty-two percent of whites agreed.[5] These contrasting attitudes, among others, are no doubt at the heart of the considerable friction between the two races at West Point.

A survey of black West Point graduates showed, also, that blacks were not so willing to recommend West Point to black youths as whites to other whites. Many of these officers also felt that race had a negative influence on their own careers. Black West Point officers have resigned from the army at a higher rate than whites, at least in the past.[6] This is disturbing in view of the fact that the new volunteer army is disproportionately black and becoming more so every day. In 1970, thirteen percent of the army was black. Today the figure is almost twenty percent, and in July 1973, thirty-five percent of all new recruits were black. The danger, of course, is that increasingly black units will be commanded by white officers. The army, in heavy competition with the civilian world, has not been very successful in attracting the intelligent, educated blacks it needs for its officer corps. Thus the Academy's desire to attract more blacks, though both are still in the throes of adjusting to each other at West Point.

Except for the Pelosi case, no occurrence at the Academy attracted so much press attention as the announcement in August 1973 that West Point had rewritten its regulations manual. This was no trivial matter. The regulations manual—the notorious 183-page (with three appendices) "Blue Book"—had been trimmed to a svelte 64 pages. But that was not all. The Academy had rewritten the book so that it was "replacing decades of accumulated restrictions with a simplified code emphasizing self-discipline." According to *The New York Times,* the revised version was "part of a new philosophy that the Academy is introducing to adapt to a changing society as well as to respond to its critics." General Knowlton was quoted as saying, "We've been manning the bastions here, hanging tight to our standards in a society that was saying there were no standards. Now we feel the pressure is off, and the first step was to redo the Blue Book."[7]

Perhaps. But a lieutenant colonel who left the Army in 1973 after a three-year tour of duty at West Point took a different view of the motivation behind the change: "I think it had to do with the new commandant. He has his eye on moving up in the army. He had to counteract the image he's had of being excessively harsh and dictatorial; his reputation as a 'flamer' [a cadet or officer who is hard on plebes] had to be modified somewhat.* With the reg changes, he looks like a progressive, flexible officer who knows how to modify the old ways in the service of the new."

Certainly, an examination of the new regulations book shows it to be a most remarkable document, a model of enlightened thinking. Though military severity flashes out here and there ("Conversation or comments audible to the extent that they preclude other persons in an audience from enjoying a motion picture or other entertainment is clear evidence of immaturity and reflects poorly on the corps of cadets"), the general tone is remarkably moderate. Under the section on professional ethics, for example, the entry reads, "Cadets are expected to adhere to the professional ethics of commissioned officers in the U.S. Army. In this regard, cadets must conduct themselves with propriety and decorum. Cadets are expected to exercise moderation in all things, in particular the consumption of alcoholic beverages, and must not engage in any activities which violate the provisions of the Uniform Code of Military Justice, state or federal laws."

Gone is the threatening tone of the 1971 version: "Cadets who drink, possess, or traffic in intoxicating liquors or who are found to be in any degree under the influence of intoxicating liquors while on or within fifteen miles of the USMA reservation will be dismissed from the Military Academy or otherwise less severely punished." The revision goes: "Cadets are not authorized to consume or possess alcoholic beverages within the limits of the USMA reservation, or in Highland Falls or Fort Montgomery. Cadets may consume alcoholic beverages at any official mixer away from West Point when offered as refreshment by the host college or as specifically authorized in these regulations." Hazing is defined the same way in both the 1971 and 1973 regulations, but the earlier version states: "Hazing is prohibited, and whenever, in the judgment of the superintendent, investigation has disclosed substantial evidence that a cadet has committed an act or acts of hazing, such cadets shall be dismissed." By contrast, the 1973 regulations say, "Hazing is prohibited. Cadets will report any instance of hazing to their immediate superior in the cadets' chain of command not involved in the incident or to their

* Shortly after Feir arrived at West Point in 1972, several officers briefing him in a meeting were heard to use words like "feel" and "believe." Feir, with some annoyance, allegedly told them, "a man *feels* with his body, *believes* in God, but in the army he'd better *know*."

tactical officer." The threat is even deferred until the last sentence: "The consequences of hazing are discussed in PARA 12.07, Regulations, USMA."

In other sections of the new regulations, more space is devoted to explaining the rationale behind certain prohibitions. A peremptory, terse order was the rule in the 1971 edition. Under gambling, for instance, the older version ran: "Cadets will not gamble or engage in any games or activities which could result in an individual's material gain or loss." The 1973 revision says, "Most forms of gambling play upon the weakness or ignorance of others; therefore, unregulated gambling generally is prohibited by local and state laws. Accordingly, cadets are not allowed to gamble. As an exception based on tradition, however, cadets may wager bathrobes on the outcomes of Navy and Air Force athletic events."

Many of the trivial sections of the old regulations have been eliminated. Cadets are no longer told that they must "not shine shoes or polish brass in the cadet locker room and will not keep cleaning equipment in their gymnasium locker," that "showers will be fully turned off after use," that "cadets will not hitchhike," or that "cadets are required to turn their bed each night and sleep in the bed, rather than sleep on top of their blankets, and use a blanket for a cover. Tactical officers and the officer-in-charge will report cadets who do not turn down their bed for 'Failure to sleep on the sheets.'"

In addition, huge sections of the old regulations have disappeared altogether. The disciplinary system is not mentioned in the 1973 revision. Nor is the honors system or the detailed regulations pertaining to accountability procedures in the academic and physical education departments, e.g., what a cadet should do if he is absent or is late to class. In sum, the new regulations, while hardly ushering in a new millennium at West Point, at least raised hopes that the Academy at last intended to encourage maturity, responsibility, and even some autonomy among its cadets.

There is now reason to believe that these hopes were premature. Contrary to appearances, a number of the regulations not found in the revised book are very much alive and in other pamphlets. Material on the disciplinary system, for example, which took up ten pages in the 1971 revision, is now to be found in the *Disciplinary System Regulations* pamphlet. Academic and physical education accounting procedures—another fourteen pages—are now to be found in the *Cadet Academic Administration Guide*. Material pertaining to inspections and formations now appears in another pamphlet, *Organization and Duties of Cadets*. More serious, however, is the manner in which West Point has undermined its own regulation changes. Punishment in various forms is "awarded" to cadets who violate the regulations. A punishment to every crime is found in a 200-page Academy pamphlet called *The Delinquency Award Guide*. When senior cadets or tactical officers wish to "award" punishment to an erring cadet, they go to the "DAG," where they are informed that not turning off the shower fully after use is a three-demerit offense, that sleeping on top of the bed instead of under the sheets warrants seven demerits, or that "expectoration in ranks" is worth five demerits. In other words, the regulations and the punishments for violation of them are as explicit and unchanged as ever; only the Blue Book has been altered. "A tac told me the DAG had been kept to insure uniformity of punishment for all members of the corps," said a cadet, "but I don't think that's the real reason."

A senior with whom I talked in December 1973 expressed pessimism about the changes: "There's been no attempt to make any real changes. The *Delinquency Award Guide* hasn't been altered at all, except to include some additional codes. When we want to see what's really allowed or not, we just go look at the DAG. I think the old regs will just carry over. Nothing will change in the long run.

"If anything," he continued, "we have fewer privileges than we had before. The new regs are supposed to put emphasis on personal judgment, but we've been told that firstclassmen can't miss Friday supper formation to escort women until second semester, that we can't be out of our rooms after 11:00 P.M. to make telephone calls, that we can't make use of facilities on post or escort after noon, and that all lights have to go off at 1:00 A.M. Before, late nights were okay; now, if you have a paper the next day, it's tough. This is the worst year I've had yet at West Point. I hope we revert to the old regs."

Perhaps, though, it is still too soon to judge what will be the effect of the new regulation changes. In March 1974 a cadet said, "The regs seem to me more open to interpretation now. Frankly, if I see a cadet do something that was forbidden under the old regs that isn't mentioned in the new, I ignore it. I've heard that less quill has been given out this year than last, and because lots of interpretations of the new regs are coming down all the time from the S-1—the adjutant—we can't be held responsible for not correcting someone for something that was in the old regs." I hope, for cadets' sake, that this cadet is right. If not, they will only have seen a public relations job, a pretense of change for the sake of appearance. Many cadets are cynical enough as it is about West Point.

Changes in the curriculum and in the military training program have evolved gradually at West Point. The elimination of mandatory chapel and the introduction of a sizable number of blacks into the corps are changes that have occurred more abruptly. Other changes that will undoubtedly alter the face of West Point are already on the horizon. If history is any guide—that the regulations seem not to have been fundamentally changed is a sobering reminder—changes inspired from within the institution will be far less dramatic than changes forced upon the Academy by the civilian world. West Point, like all bureaucratic structures, has a remarkable ability to preserve its internal homeostasis; changes in one place usually cause compensatory changes elsewhere. Whether dramatic changes inspired from without are more effective and lasting than those prompted from within the Point, however, remains to be seen.

A dialogue is currently taking place between officers and cadets at West Point. Both are concerned about two issues. First is the tendency of West Point to develop leadership skills only in those who have already shown leadership ability. As one cadet said, "The system promotes a cycle. If you show leadership ability early, you get high ratings on aptitude, which leads to higher positions of leadership. And the higher you are in the chain of command, the better aptitude ratings you get. Training here overemphasizes the best at the expense of cadets who aren't so good." In other words, West Point tends to select rather than develop leaders. The second concern is that cadets, especially those of proven leadership ability, are often unable to assume the roles and responsibilities of officers (which they soon will be), even when they are seniors at the Academy.

One proposal attends to both problems: seniors picked as company or battalion captains would serve in those positions two out of the three school terms. They would then move out of their companies into separate quarters and would return only to lead formations or inspect their companies. The rest of the time, in addition to their usual academic duties, they would work in staff positions, in closer contact with officers, on projects such as regulation changes. As the captains moved out, other seniors who had previously served as sergeants or platoon leaders would assume the vacated captaincy slots and get their opportunity to exercise leadership.

In December 1972 a committee which included Frederick R. Kappel, president of American Telephone and Telegraph Company; Frank Rose, National Chairman for Planning and Development of the Salk Institute; General Charles H. Bonesteel, III, formerly Commander in Chief, United Nations Forces in Korea, and Commanding General, Eighth United States Army; Roy Lamson, a professor of English at the Massachusetts Institute of Technology; and Colonel Edward A. Saunders, head of the physics department at West Point, published the findings of a months-long examination of the Academy's education objectives, its curriculum, its faculty, its methods of instruction, and its leadership program. This report, known as the Kappel report, approved of West Point's educational objectives. The committee agreed that the Academy's reluctance to institute academic majors was sound, that the Academy should retain its present broad curriculum, that the predominantly military faculty be retained and the present selection and education policies be continued ("We find that present selection and education policies have been successful in maintaining an excellent faculty at USMA.").

On the other hand, the committee recommended that the mathematics load for freshmen be reduced, that all cadets be allowed to take fewer courses, and that the Academy give consideration to increasing use of visiting professors. The strongest recommendations, though couched in exquisitely tactful language, were aimed at the military training program. The Kappel report stated, "We find some indication that the Leadership Development Program may be excessively concerned with detail and that it tends to overload the cadets with functions which are not clearly essential. There is also evidence that the system could be more effective in developing a sense of self-discipline and individual responsibility in a cadet." The committee recommended that the Academy "analyze" its entire training program, including its leadership evaluation system, the structure of tasks and positions within the corps, the relationships between cadets and tactical officers, the system of rewards for cadets, and the regulations under which cadets live.

To its further credit, the committee noted that a cadet's time is overscheduled, "which sometimes forces him to an expedient slighting of one or more of these multiple demands." Finally, the committee voiced concern that the average cadet was not well informed concerning either the Academy's program or current issues involving the army and the Military Academy. "This lack of information and understanding extends across curriculum, local problems and issues facing the Academy, and national events affecting the army."[8]

Whether the Kappel report will have any impact upon the Academy is an open question. It does appear that the Academy is moving to implement the committee's recommendations. On March 19, 1974, General Knowlton made a closed-circuit television appearance before the corps of cadets. A senior summarized the performance: "General Knowlton described the members of a committee who had recently studied the Academy's curriculum and told us what their credentials were. He said that the curriculum, while steeped in tradition, continually undergoes scrutiny. The latest research indicates that cadets are overloaded. The big news, though, is that all academic departments have been ordered to cut their course time ten percent. For example, plebes attend math classes six days a week now; next year they'll have Wednesday off. The foreign language department will cut their course ten percent during cadets' second year. The crux of Knowlton's speech was to inform us of what policies will be implemented next year and of the rationale behind those decisions. I think it's a step to make the corps more aware of processes that shape their future."

Rumors have circulated within the corps that the Academy will alter Beast Barracks in at least one significant way in 1974: new cadets will no longer be required to sit at rigid attention throughout the meal. They will be allowed to eat their food without harassment. "But the Academy wants to take up slack in other ways," a cadet warned. "Last summer upperclassmen were permitted to make on-the-spot corrections on new cadets, but unless they were in the new cadets' chain of command, they weren't allowed to give them special inspections [inspections lasting more than three minutes]. The policy was started to prevent an upper-classman from running a new cadet out of the corps if he wanted to. But lots of upperclassmen felt that this kept them from correcting new cadets properly. There's been talk about doing away with that rule next summer. Lots of cadets are against Beast Barracks changes. They think a tough Beast Barracks is a tradition that has worked in the past, and if we change anything it might produce poor officers. Cadets are also very egocentric. It's like saying, 'Look what a good cadet I am because I went through it.' "

While giving talks on psychological stress to seniors on the new cadet detail in 1971, I noticed the same reaction from upperclassmen. Several cadets were annoyed when I suggested there were ways to lessen the pressure on a man if he showed unmistakable signs of psychological stress. One man even said, "We had to go through it. Why shouldn't they?" Part of this response, I suppose, was the automatic military re-action: any easing of training leads to "softness." Other cadets were probably worried that the fraternity would not be so elite if the initiation rites were tampered with. I also suspect that many upperclassmen who have gone through Beast Barracks and plebe year greatly resent the man-ner in which they themselves were treated. Harassing new cadets is a form of displacement of this resentment.

West Point is likely to be confronted with accelerating demands for change in at least two other areas very soon: its honor board proceedings and its all-male admissions policy. In fact, assaults on the honor system have already begun. In May 1973, six West Point cadets accused of cheating on a physics examination challenged as unconstitutional the procedure by which the cadet honor committee administered the honor code. They claimed that during their hearings they were denied the right to consult a lawyer, were never told who was testifying against them, were not informed of the nature of the testimony, and were never told of their right to remain silent. In addition, they were given as little as three hours to prepare a defense after they had been accused. One of the six cadets said that he had asked an honor committee member why he had been found in violation of the code. He was told, "The point is, we don't need concrete proof that you cheated. We don't have to base our decision on that. It can be a feeling among us."

The suit was subsequently lost, but the judge did not comment directly on the cadets' allegations. Instead, he dismissed their charges by ruling that deficiencies in the honor committee's procedures could not form the basis of a lawsuit because every cadet charged with a violation had the right to appeal his case to a board of officers. The judge said that the cadets would have had to show that the officers' board proceedings were unconstitutional in order to prove their case.[9]

In December 1973 an Air Force Academy graduate, Captain Michael T. Rose, charged that all service academies allowed extensive violations of constitutional guarantees. His report, a two-year study which examined the codes of West Point and the Air Force Academy more closely than the codes at the other academies, was completed while he was a law student at New York University. The report states that an alleged disregard for "legal technicalities" fosters a contempt for the law that leads to the justification of "undesirable military practices," and that the exaggerated sense of moral superiority engendered by the code enables the nation's highest ranking officers, eighty-five percent of whom in the army and navy are Academy graduates, "to accept a My Lai massacre or fourteen months of secret bombing of Cambodia as protection of the nation." Rose cited the practice of silencing as "a blatant example of cadets learning that no matter what the law requires, if their personal value judgment dictates otherwise, the law can be ignored." He suggested that violations of an individual's rights could only be solved by "codification of the honor and ethics codes, having honor boards bound by precedents, and adopting procedures developed by qualified lawyers in accordance with constitutional requirements." He added that Congress must institute these moves "because the academies, highly inbred, are unable to do so." As General Omar N. Bradley (class of 1915) said when he accepted the Academy's Sylvanus Thayer Award in May 1973, "I'd hate to see anything happen through the courts or anything else that could lessen honor training."

It is also probable, in light of trends at the other service academies, that West Point will soon admit its first women. In January 1974, Representative William Hungate, Democrat of Missouri, nominated an eighteen-year-old girl to the Air Force Academy. A month later, two girls were nominated for entrance into the class of 1978 at the Merchant Marine Academy shortly after the academy announced that it would accept women. Both of the girls—one an honor student as well as the second-best baton twirler in New York State, the other also an excellent student with an outstanding record as a gymnast—reportedly "hoped the Academy would be like any other college for them." A West Point cadet said, "The Academy has a contingency plan for admitting women. Uniforms and barracks are ready for them." Perhaps as a prelude to this new era, West Point in 1973 invited its first woman officer, a lieutenant, to teach a course in environment in the Department of Earth, Space, and Graphic Sciences.

The reaction of West Pointers to the possible admission of women is predictably mixed. A 1972 graduate said, "I think it would be kind of nice, but it's weird. If more than fifteen women came to West Point—they'll just be a novelty if there're fewer than that—it might make West Point more like a civilian campus. Lots of guys at West Point don't have a girl as a friend. They get to know girls as dates and lovers, but never as friends; they might get to know what girls are about—some insight—if they were around. They also wouldn't have to truck in that many girls for dates. What I don't know is how the Academy will tear them down and build them up like they did us during Beast Barracks. Or what they'll do about haircuts."

A current senior at the Academy was more negative about the prospect of women coming to West Point. "Personally, I don't like it. Women aren't qualified for the combat arms branches. They can't get up and run up a hill and take it under fire. Why should the government spend seventy thousand dollars to turn out a woman to go into the quartermaster corps?"

This same cadet was also enlightening about the politics which so far have helped keep women away from the Academy. "At the beginning of the year [fall 1973] a small number of cadets were going to be allowed to go into each of the noncombat arms branches—e.g., the finance corps, the signal corps, the medical service corps—for the first time. Before, only cadets who were medically disqualified for combat arms were allowed to join the noncombat branches. The Secretary of the Army found out about the plan and put a stop to it immediately. He realized it would have shot a hole in the argument that women shouldn't be admitted to West Point because they couldn't qualify for combat arms. If cadets were going into noncombat arms branches, why shouldn't women, too?"

But whether or not West Point admits women as cadets or invites them to come as instructors, they will have still another problem to consider. In July 1973 a Federal judge in Brooklyn struck down as unconstitutional a regulation of the Merchant Marine Academy that prohibited marriage of midshipmen. In a thirty-six-page opinion, Judge John R. Bartels wrote, "The fatal vice of the regulation is the sweeping, advance determination that no married student, regardless of age, maturity, or circumstance, can be accepted, or if unwittingly accepted, must be expelled from the Academy simply because he is married. The conclusive presumption that all married cadets will perform poorer than single cadets cannot be accepted upon the record before the court." The judge also noted that he found no marriage prohibitions among military academies of other nations, including Britain, Norway, Japan, Nationalist China, and West Germany. The French, he observed, permit a student to marry in the second year and, "in a delicate situation," in the first year. Seven months later, four midshipmen at the Merchant Marine Academy were married, and a reporter noted, "Theoretically and legally, a married couple could now attend the academy together."[10] There is no doubt that the Bartels decision, if upheld by higher courts, will affect the no-marriage policies at the other service academies, including West Point.

No doubt many of these changes will create a clamorous protest from West Pointers who fear that venerable Academy traditions are being destroyed. As one cadet commented, "I've heard old grads say, 'West Point's what made this country great. You don't want to mess with the mortar that's held the nation together for two hundred years.' "

The Academy, of course, is in a dilemma. It is unsure of itself, uncertain which way to turn. On the one hand, its strong conservative constituency demands that changes be made slowly, if at all. It implores the Academy to recognize that its very structure and rigidity may well be one of its greatest strengths. As General Knowlton said in 1970, "[The cadet] wants a system where virtue is rewarded and error is punished. He likes a structured environment. He's a man who believes in absolute values and will tend to reject the situational morality of today. He believes there are unchanging things in this changing world."

On the other hand, the Academy cannot disobey Federal court rulings which have eliminated its chapel policy and which threaten to alter its honor system. Nor can it disregard the precedents currently being set at other service academies.

The challenges to West Point's honor board proceedings may well alter a long tradition, but it is difficult to imagine that the replacement of those procedures by the Universal Code of Military Justice could be disadvantageous to the prospective West Point officer. It is, after all, the legal system he must adhere to after he graduates. And though West Point would no doubt look different with female cadets marching and walking about, it is hard to imagine that women would disrupt the teaching of order and discipline. Their presence, in fact, might have the opposite effect. Cadets could hardly be more preoccupied with women than they presently are. The opportunity to see them every day might well reduce their obsession and allow them to see women more realistically, as human beings rather than as objects of pleasure or veneration. At Stanford University, as at most civilian colleges, men and women found that the coeducational experience deflated the mystique that inevitably flourishes when males and females live isolated from each other, a mystique which often accentuates difficulties in relations between them. Experimenting with coeducational living, Stanford discovered that destruction of property and rule-breaking became much less frequent in dormitories that were sexually integrated. It may be that women, not music, soothe the savage breast.

Proposals for change which emanate from within the Academy more often resemble Brownian movement than linear progression. In writing a foreword to Stephen Ambrose's book *Duty, Honor, Country* in 1966, Dwight Eisenhower, who had graduated from West Point fifty-one years earlier, admitted he was "amazed at how little some things change." But, like it or not, the demand for constitutional rights and equality for women and blacks—which are largely changes forced upon West Point from the outside—are issues the Academy cannot avoid. These are the issues of the real world and, more immediately for cadets, of the army itself. If West Point is to remain true to its word, to its avowed intention of producing a multidimensional officer able to move with ease, understanding, and competence in different worlds, cadets must face and live with these issues while they are still at West Point. Whatever the changes, internal or external, it will be interesting to see if and in what ways they fundamentally alter this most conservative of military academies.

12

CONCLUSIONS

Just how valuable is West Point training? How well does it serve the men it produces?

For a man interested in an army career, West Point is the "premier card punch," as one officer rather cynically expressed it. Though West Pointers make up fewer than five percent of all army officers, most of its generals are Academy graduates. However, this may say more about the preference West Point graduates have for each other than it does about the effectiveness of West Point training. A major in the mathematics department, in a statement more remarkable for its feeling than for its logic, said, "Some people are jealous because West Pointers make it to the top more often than nongraduates. Why not? If they didn't make it more often, what point would there be to having a military academy?" West Pointers are well aware of the W.P.P.A.—the West Point Protective Association, an informal organization which critics, usually non-West Pointers, claim exists to protect and foster the interests of Academy graduates—but there is very little talk about it at West Point. One officer, however, remarked, "The W.P.P.A. isn't a big conscious thing, but it exists. You get to know people over time, starting here. Where West Point really helps, though—and this is one of the

few places you can see it directly—is in your choice of a first assignment. Being from the Academy, you have a better chance than an
R.O.T.C. guy of getting into a good unit in, say, Germany. And if you
do a good job for people on their way up in the army, you go right up
with them."

"On AOT last summer," recalled a cadet, "a non-West Point officer
complained to me that West Pointers look out for each other and themselves. He got annoyed when I said, 'I hope they do.' "

Ward Just, in his book *Military Men,* has written:

There is a concomitant (besides the W.P.P.A.) to the advantages of an
Academy education: familiarity with the system from age eighteen and the
astonishingly strong ties that old grads have with West Point and hence with
cadets. The West Point ring will not indefinitely protect a real maverick or a
definite incompetent, but in the great middle range of army officers, it does
a good deal more than a Harvard Law School degree does for an ambitious
corporate lawyer.[1]

The relationship between a man's performance as a cadet and his subsequent career as an officer remains less clear, however. It has been
established, for example, that a cadet's academic standing correlates
very poorly, if at all, with the rank he attains as an officer. One study
showed, in fact, that the best predictor of a man's becoming a general in
the army was an outstanding athletic career at the Academy.[2]

Historical biographies of unusual men do not prove any relationship
between a young man's record as a cadet and his later career as an
officer. Bruce Catton's biography of Grant in no way substantiates a
favorable influence of West Point on Grant's career. Grant went to
West Point at his father's insistence. He was a mediocre student who
excelled only in mathematics and horsemanship, and he was often at
odds with West Point tradition and discipline. "Never could the army
quite eradicate his faint air of slouchiness, or take the little stoop out
of his shoulders, or induce him to pay more attention to the spit and
polish aspect of soldiering."[3]

During his years at West Point, Grant fervently wished Congress
would abolish the Military Academy; after graduation he hoped to
obtain a permanent position as a professor in some respectable college.
At graduation, he stood 156th in order of conduct out of a total of 223
cadets in the corps. There is evidence that he never read a book on
military strategy. As he himself said, "I don't underrate the value of
military knowledge, but if men make war in slavish observance of rules,
they will fail."[4]

134

Grant, in fact, harbored a lifelong dislike of the military profession. Although he retained an immense and curious (in view of his own troubles with it) respect for West Point's disciplinary system throughout his life, Grant did not have fond memories of the Academy: "I hear army men say their happiest days were at West Point. I never had that experience. The most trying days of my life were those I spent there, and I never recall them with pleasure."[5]

General William T. Sherman, "the most original genius of the American Civil War," entered the Academy at sixteen, two years after his foster father peremptorily told him "to prepare for West Point." He had a superb academic record and graduated sixth out of forty-three cadets in his class, but he did not consider himself a good cadet; he remained a private throughout his four years at West Point. "Then, as now," he later wrote, "neatness in dress and form, with a strict conformity to the rules, were the qualifications for office, and I suppose I was found not to excel in any of these."[6] By graduation, he had amassed an average of 150 demerits a year.

This brilliant, high-principled, restless man sought to resign his commission in the army when he was twenty-nine but was turned down. He finally left in 1853 at the age of thirty-three. After working as a banker for four years, he assumed the superintendency of the Louisiana State Military College (later Louisiana State University) and stayed on until the outbreak of the Civil War in 1861, when he returned north and shortly afterward was appointed a colonel in a regular army regiment.

General John J. Pershing graduated thirtieth out of seventy-seven cadets in his West Point class of 1886. He had no intention of making the army his career, having always hoped to become a lawyer. Later he received his law degree from the University of Nebraska, but he went to West Point because it provided a good education at government expense.

As the oldest man in his West Point class, Pershing showed unmistakable leadership qualities. He was elected president of his class and named first captain his senior year. Always a stern, though fair, disciplinarian, he ranked only seventeenth in his class in discipline. Pershing had a curious trait. He was forever tardy; he could never make formations on time.

Returning later to West Point as a tactical officer, Pershing was decidedly unpopular. The cadets in his company "chaffed under Pershing's relentless demands and frequently abrasive personality."[7] A chance meeting with Theodore Roosevelt at Madison Square Garden in 1897 was probably the turning point in Pershing's career. The two men took to each other immediately; their common interest was the American West. Roosevelt was later instrumental in securing a key promotion for Pershing.

MacArthur and Eisenhower were both generals of the army, but their backgrounds and West Point careers were very dissimilar. MacArthur's father had attained the rank of lieutenant general in the army by the time he died in 1912. MacArthur formed an unusually close relationship with his son, who admired him not only as a parent but also as his ideal of a military leader.

After four years at the West Texas Military Academy in San Antonio, where he won all the prizes, MacArthur made two unsuccessful attempts to secure a presidential nomination to West Point. He finally took the examination for the Academy in 1898, placed first, and entered in the class of 1903. He suffered a fearful hazing as a plebe because of his father's reputation but went on to achieve a brilliant record at West Point. He was named first captain his senior year and graduated first in his class, compiling one of the highest scores on the general merit role in the history of the Academy. MacArthur's roommate commented that he was "one of the hardest workers I have ever known," and "his energy was directed to the attainment of . . . number one in his class."[8]

Dwight Eisenhower had no idea what he wished to do when he finished high school. He worked in the town creamery until he met another young man who at the time was preparing for Annapolis. Eisenhower decided that he needed a definite goal. He took the exams for both Annapolis and West Point but entered the Academy because he was too old—twenty—to attend Annapolis.

He, like Grant, was a mediocre student who had trouble with the disciplinary system. "I was," he said, "in matters of discipline far from a good cadet." One of his biographers records: "Eisenhower was a roughneck. He broke the rules just as often as he dared."[9] His record confirms this: he was "awarded" exactly one hundred demerits his senior year and stood 125th out of 168 cadets in his class that year. He never held a rank higher than color sergeant. His real interest was football, which he played well until he injured his knee during his second year. He became a cheerleader his third year and coach of the plebe football team his senior year. Though popular and independent, he seemed to have impressed no one as a future commander. He was called "average," "nothing outstanding."

Brigadier General Sam Walker, recently commandant of cadets at West Point, has said that before a man can lead, he must be able to follow, follow absolutely. But Basil Liddell Hart has written, "Even among great scholars there is no more unhistorical fallacy than that, in order to command, you must learn to obey. A more temperamentally insubordinate lot than outstanding soldiers and sailors of the past could scarcely be found."[10] The careers of Grant, Sherman, and Eisenhower bear out this contention. All of them, while certainly not insubordinate, had their difficulties with discipline at the Academy, and the marks they received in it were inversely related to their later success.

What all five famous West Point graduates shared, however, were relatively secure childhoods in stable families, which instilled in them a sense of responsibility, obligation, and self-discipline. These attributes were well established in their characters long before they entered West Point. The ability of the Academy to enroll such men, for whatever reason—Grant and Sherman because their fathers ordered them to West Point, Pershing because he wanted an education and could not afford it elsewhere, MacArthur because of the example of his father, Eisenhower because it provided a structure for his life—is undoubtedly at the heart of West Point's success in "producing" competent military leaders even while it discourages initiative and responsibility. Liddell Hart wondered, in fact, whether it was because of or in spite of the Academy that some West Point graduates became successful leaders:

As for the argument that this system of "education" has produced many great leaders, its validity depends upon the complementary question whether they have become great because or in spite of the system. The criterion "by their fruits ye shall know them" is fallacious if there is only one tree which grows fruit, one way by which a profession can be entered. The real question, then, is why the blight was felt by so many blossoms and whether even the best fruits would not have been better if they had escaped the blight as blossoms.[11]

A 1963 West Point graduate said, "The only responsibility you have in order to get promoted is to follow the regs. For a guy to take responsibility, he has to go out and beat the system. West Point develops good followers. It says it develops good leaders, but leaders have to see possibilities. The minute you give cadets something to do on their own, they ask, 'Where are the guidelines?'

"West Pointers show no initiative; they're yes-men. As a company commander, I had O.C.S., R.O.T.C., and West Point lieutenants working under me. The O.C.S. guys weren't much; they got through by being nice. The R.O.T.C. guys were hard chargers; they looked for ways to be good leaders. The West Pointers waited for people to tell them what

137

to do. They'd say, 'Tell me how the guy before me did it, and I'll do it better.' "

As a matter of perspective, one should remember that cadets' leadership training comes after graduation as well as before. In a succession of assignments to infantry school, for example, advanced weapons training school, service in an infantry company, Command and General Staff College, and perhaps the War College, officers have many opportunities to demonstrate their suitability for higher leadership positions within a system dominated by other West Pointers. To speak of West Point as "producing" leaders is inaccurate. The Academy is only one, albeit an important, way station in an army career. The kind of leader and/or the rank a man attains in the army depends as much on chance, "old-boy" ties with other Academy graduates, his experiences after graduation, and the kind of person he was before he entered West Point as anything that inherently goes on at the Academy, though West Point without doubt teaches its cadets the prerequisites for success in today's army.

If West Point does not make leaders, what kind of men does it make? What do West Pointers really learn at the Academy? One source of testimony comes from West Pointers teaching at the Academy. A 1962 graduate in the dean's office said, "There are three important things I learned here. One was honesty. At the college I went to before I came to West Point, I cheated by taking exams and writing papers for people and by turning in the same paper three times. The honor code makes you do your own work. Second, if you survive West Point, you become confident in your ability to survive in any environment. After four years of being told how tough it is, you say to yourself at the end, 'God damn it, I'm good.' Third, you form good friendships here, friendships you know you'll have all your life. You've been through a lot of the same things together and you don't forget that."

A second officer, a 1961 graduate assigned to the mathematics department, commented, "I learned what the concept of duty—responsibility to superiors and subordinates—meant, especially during my plebe year as a mail carrier. You learn to do your job when you know people are depending on you. I also learned about 'cooperate and graduate,' which taught me a lot about getting along in the army."

A tactical officer observed, "You learn how to get along with all kinds of people here, even people you don't like. You learn to adjust to all kinds of situations and you learn the importance of teamwork."

A major in the Department of Military Psychology and Leadership said, "There are two aspects of West Point that stand out, one positive and one negative. The positive has to do with the development of the urge to do something worthwhile in terms of service, to do something that contributes, to serve. West Point promotes the nobility of sacrifice. Unfortunately, the system also develops a passive attitude toward life which is reinforced in the army. Then, when you want to be independent, and that's where I'm at now, it's tough."

Cadets give much the same report, usually emphasizing a sense of achievement, friendships, self-discipline, appreciation of physical fitness, and the opportunity to travel. In a questionnaire given to the West Point graduating class of 1972, seniors were asked what they felt was most useful to them about their training and education at West Point. Their answers confirmed what the officers had said. Cadets cited, in order, the ability to tolerate pressure, the ability to use time efficiently, the ability to communicate and work well with people, and the development of self-confidence.

Poise, pride, self-confidence, self-discipline characterize the positive aspects of West Point for officers and cadets alike. No doubt hundreds of recitations in class, countless parades, and hours of instruction in "command voice" give a man a certain sheen, a certain presence. Pride and self-confidence are the result of achievement and survival through four tough and demanding years. As a 1972 graduate said, "I had a tremendous sense of accomplishment when I graduated. I wanted to prove to myself that I could get through West Point, and I wanted to show my parents I could put myself through school. Graduation was the culmination of both of those desires." The self-discipline cadets speak of is the ability to finish one's tasks and work efficiently in the face of great demands on one's time.

Who could deny that increased self-confidence, discipline, the capacity to use one's time efficiently, the ability to adapt to stressful circumstances, a sense of responsibility toward one another, and the ability to work effectively with other men are all highly desirable for a successful military leader? These are important aspects of West Point training which should not be underrated. They are also the qualities every army officer—the battlefield commander as well as the military manager —must possess.

But the development of these attributes takes place at a high cost. The same training which inculcates those traits transmits and reinforces other traits as well. At least part of this immensely skewed development can be defended on the grounds of West Point's harsh obligation to create fighting men who excel not in the arts of peace but in the arts of war. Personal characteristics not generally admired within civilian society can have their place in the military world. Suspiciousness and

139

even paranoia may be appropriate traits in a man whose professional responsibility lies in calculating the threats to his nation's security, a man who must always estimate the capabilities of the enemy. Dependence, too, while often depreciated in the civilian world, has its purpose within the military society. An officer who feels insecure in the outside world is a man obviously more attached to his army world, and a man who will make a more reliable, pliant team member than a person who feels more independent of the institution. Unfortunately, however, other lessons learned and reinforced at the Academy serve to undermine the officer corps, the army and the military mission of serving the country.

By isolating its cadets—physically, psychologically, intellectually, and morally—West Point deprives them of any real understanding of the outside world. And without that understanding of the larger society, how can they possibly serve it effectively? For instance, the rigidity of the honor code leads cadets to regard justice as the arbitrary, unthinking application of a law or principle. The effects of this attitude on the society which cadets ultimately reenter is incalculable, but it undeniably results in a blunting of moral judgment.

Examples abound. At the time it was revealed in the press that the army had been spying on civilians, a West Point instructor asked, "What does it matter who's spying on you if you're not doing anything wrong?" Another instructor, a West Point graduate teaching in the social sciences department, objected to carrying on the war in Vietnam as immoral "after we saw we couldn't win it in 1968." A third graduate, also teaching in social sciences, told me of the great numbers of Americans killed in the Civil War, the First World War, and the Second World War. Then, referring to the smaller number of Americans dead in Vietnam and the sophisticated weaponry used against the enemy, he concluded, "War is more civilized now." When the dual system of reporting bombing raids in Cambodia was being investigated by Congress in July 1973, a West Pointer, General George S. Brown, class of 1941 and now Chairman of the Joint Chiefs of Staff, declared that the falsification of records could not be considered lying if "those who ordered and planned the raids would not have been deceived."[12]

It is also impossible to develop a genuine moral sense when duty is as narrowly defined as it is at West Point. One cadet, a high-ranking senior in the chain of command in 1974, admitted, "The biggest thing I've learned at West Point is the meaning of duty: the willingness to carry out orders and do a good job according to someone else's decision. If my CO asks me to do something, I do it well, even if it conflicts with my personal views. I came here as a liberal, someone who was willing to stand up for my rights, but now I can do something I don't believe in just because my CO does. It's something I'm not especially proud of, but it's happened.

"You do things because it's more important to perform your duty than to stand up for what you believe. Oh, I can argue with my CO, but once he makes up his mind, I'll carry out his instructions to the limit. It used to be hard, but it's not anymore."

The lesson is permanent, it would appear. To hear a powerful, high-ranking West Point general admit that he would sacrifice his integrity to obedience is even more dramatic, as one did in July 1973. General Earle G. Wheeler, a 1931 West Point graduate and former Chairman of the Joint Chiefs of Staff, was questioned by Congress about the falsification of bombing records in Cambodia. He "expressed horror at the falsification of the records, but said that if the President had ordered him to falsify them, 'I would have done it.' "[13]

Many of those who manage to resist this moral immunization—the disenchanted, the apathetic, the cynical, and perhaps the enlightened—simply serve their five years in the army and resign. There is evidence that their numbers are increasing. In the class of 1956, for example, only fifteen percent had resigned after five years. In contrast, the class of 1966 had lost twenty-eight percent of its members after five years. By March 1972, the same class had lost thirty-two percent of its members.[14]

In 1971, a study published by the Academy's Office of Institutional Research showed that forty-six percent of the cadets who made up the class of 1971 planned to leave the army before twenty years were up. Forty percent of the class said that they would not attend West Point if they had to make the choice again. In contrast, only twelve percent of members of the 1957 class said they would not come to West Point if they were given a second chance.[15]

No doubt these resignations and changes of attitude reflect disillusionment with Vietnam, the army's role there, a general disaffection with governmental institutions, and a host of other personal reasons. But in many cases they reflect a man's unwillingness to sacrifice integrity to the demands of the institution. When asked by *New York Times* reporter Seymour Hersh in 1972 why he had decided to resign from the army, Josiah Bunting said, "Mainly I think it's dissatisfaction with the whole deadly career game. You know you've got to play it safe for a long period of your life, so you can become a general—and by the time you reach it, you're a completely different person."[16]

"We're always debating with each other," said a 1960 West Point

141

graduate, "about which stage of your career they suck your brains out." Basil Liddell Hart observed the way that "ambitious officers when they came in sight of promotion to the generals' list would decide they would bottle up their thoughts and ideas, as a safety precaution, until they reached the top and could put these ideas into practice. Unfortunately, the usual result, after years of such self-repression for the sake of their ambition, was that when the bottle was eventually uncorked, the contents had evaporated."[17] I need hardly add what it means for West Point, the army, and the country when cadets and officers who have retained a sense of integrity resign their commissions.

Although West Point can hardly be held responsible for all the problems and failures of the army, the fact remains that West Pointers do control the army. The malaise which afflicts it today no doubt owes a great deal to its West Point core. It should come as no great surprise that the lessons of the Academy are now the problems of the army: obedience above integrity, a reckless overemphasis on competition, an unwillingness to admit failure and errors for fear of looking bad and jeopardizing one's chance for promotion, unrealistic optimism, powerful in-group loyalties whose main function is to protect members of the group, a disregard for rights of the individual, and isolation from current problems. Indeed, there is a parallel list of words and phrases in the army for the West Point precepts: "cover-up," "can-do," "careerism," "favoritism," and "elitism."

Several critics of West Point and the army have called for radical alterations in the Military Academy as it now exists. One of them, Edward L. King, a former lieutenant colonel and a non-West Pointer, holds the Academy responsible for the current demoralization in the army. He maintains that its leadership clique has invariably insulated the army from reform. He proposes that the Academy close its doors as an undergraduate institution and that army officers be obtained at much less cost from civilian scholarship programs and R.O.T.C. West Point could then be transformed into the senior-school center of the army, where the Army War College, the Command and General Staff College, the senior NCO school, and the army museum and library would all be located.

Bruce Galloway, a former intelligence officer in the army, and Robert Johnson, a 1965 graduate of the Academy, proposed in their 1973 book on West Point that the Academy be abolished and each of the four regiments be sent to selected locations throughout the country so that cadets will feel more comfortable with the people they serve.

In 1969, a critical study of all the major service academies was published by J. Arthur Heise, a former librarian at the Air Force Academy while on active duty from 1962 to 1965.[18] All Academy papers went through his hands on their way to the Air Force archives. Heise did not call for abolition of the service academies—"To call simply for their abolition is an exercise in futility in view of the deep roots the Academies have in the American scene." Heise concluded that the academies must be substantially changed, and proposed that a special presidential commission be appointed to study and define the attributes that the Academy graduates of today and of the foreseeable future would have to possess. Heise was especially concerned that such a commission rectify the Academies' tendencies to teach its cadets and midshipmen simple, dichotomous, black-and-white methods of thinking and problem-solving. "Their graduates," Heise wrote, "must not only be ready to go into battle, they must also be equipped with the ability to shift mental gears to the larger spectrum of grays that will confront them during *most* of their career." His further recommendations were more radical than, but strongly reminiscent of, the later Kappel report: that the faculty be upgraded, that authority and responsibility for internal educational policies and related matters be placed in the hands of the faculty at large and not confined to the academic board, that a more critical and active board of visitors be instituted to see that the schools' overall goals were met, that the fourth-class system be altered, and—perhaps the strongest of the proposals—that cadets and midshipmen at all the academies be brought back into contact with American life and their parent services.

In 1974 a cadet said to me, "We were once told that the stream—meaning the army—is only as pure as its source. West Point strives to keep the source pure. It knows that the whole stream will end up polluted farther down, but that's a better state of affairs than if it's polluted at its source." "Polluted" is too strong a word to apply to the source—West Point—but there can be little doubt that it is not so pristine as the Academy would have the public believe.

The headwaters—the Academy's ethical system—is an immensely complicated problem for which no easy answers exist. A system that works so poorly ought to be modified, at the very least. To insist rigidly upon an honor system which most cadets know is unenforceable, unrealistic, and unreasonable only leaves the institution open to disrespect and charges of hypocrisy. The Naval Academy does not have an honor code. Instead, it has an honor concept that stipulates that midshipmen will be "truthful, trustworthy, honest and forthright at all times and under all circumstances." Under the Annapolis concept, one midshipman is not required to report another for, say, cheating on an exam. As a midshipman explained to a newspaper reporter in March 1974, "If an army cadet fails to report a violation, he has violated the code himself and may be dismissed. But a navy officer is expected to show more moral maturity—to do the right thing because it's right, not for fear of punishment. We must exercise our moral judgment in each case. We can report a violation through the honor system. We can counsel the man, or we can do neither. If we do nothing, while we have not committed an honor violation, we would have failed in our responsibility to the honor concept and to the brigade of midshipmen."[19]

One consequence of the navy's honor concept is that a much smaller number of midshipmen are lost to the Naval Academy each year because of honor infractions. And, to my knowledge, there is no proof that Annapolis graduates are less honorable than West Pointers.

My principal intent, however, is not to prescribe for the Academy. The military profession is no more exempt than any other organized occupation from Bernard Shaw's observation that all professions are conspiracies against the laity. My purpose, rather, has been to pierce the armor of misinformation, tradition, and myth that surrounds West Point so that we may inspect the Academy more thoroughly and decide for ourselves if what goes on at one of our most venerable institutions is really in our best interest.

NOTES

Chapter One: The World of West Point

1. Gore Vidal, "West Point and the Third Loyalty," *The New York Review of Books,* October 18, 1973, p. 22.
2. D. D. Eisenhower, "This Country Needs Universal Military Training," *Reader's Digest,* September 1966, pp. 49–55.

Chapter Two: The West Point Candidate

1. J. W. Houston, J. M. Fabian, and D. L. Greco, *Characteristics of the Class of 1975* (West Point, New York: U.S. Military Academy, Office of Institutional Research, 1971), pp. 1–16.
2. *A Study of the Programs of the United States Military Academy* (West Point, New York: U.S. Military Academy, 1972), pp. 97–98.
3. L. I. Radway, "Recent Trends at American Service Academies," in C. C. Moskos, *Public Opinion and the Military Establishment* (Beverly Hills, California: Sage Publications, 1971), p. 5.
4. Houston et al., *op. cit.,* p. 26.
5. A. D. Wise, *A Comparison of New Cadets at U.S.M.A. with Entering Freshmen at Other Colleges* (West Point, New York: U.S. Military Academy, Office of Institutional Research, 1969), p. 8.
6. Houston et al., *op. cit.,* pp. 14–15.

7. J. W. Houston and D. L. Greco, *Report of New Cadet Barracks Questionnaire, Class of 1974* (West Point, New York: U.S. Military Academy, Office of Institutional Research, 1971), p. 8.

8. Wise, *op. cit.,* p. 10.

9. Radway, *op. cit.,* p. 5; Houston and Greco, *op. cit.,* p. 9.

10. Wise, *op. cit.,* p. 13.

11. *Ibid.,* p. 15.

12. J. W. Houston, *Some Attitudes of Selected Subsamples of Cadets, Class of 1974* (West Point, New York: U.S. Military Academy, Office of Institutional Research, 1971) p. 2; W. E. Hecox, *A Comparison of New Cadets at U.S.M.A. with Entering Freshmen at Other Colleges, Class of 1973* (West Point, New York: U.S. Military Academy, Office of Institutional Research, 1970), cited in Radway, p. 7.

13. Wise, *op. cit.,* p. 22.

14. Houston and Greco, *op. cit.,* pp. 37–38.

15. *Ibid.,* pp. 24, 29–30, 39.

16. J. R. Moskin, "Who Would Ever Go to West Point Today?" *Look* (October 6, 1970) p. 36; unpublished statistics, Office of Institutional Research, West Point, New York, 1972.

17. Houston and Greco, *op. cit.,* p. 42.

18. J. W. Houston, *Trends in Admission Variables* (West Point, New York: U.S. Military Academy, Office of Institutional Research, 1971) pp. 28–29; Wise, *op. cit.,* p. 9; Houston and Greco, *op. cit.,* p. 4.

19. J. W. Houston, *Trends in Admission Variables, op. cit.,* p. 13–14.

Chapter Three: Beast Barracks

1. L. Tiger and R. Fox, *The Imperial Animal* (New York: Holt, Rinehart and Winston, 1971) pp. 158–159.

2. P. Bourne, "The Military and the Individual," in J. Finn (ed.), *Conscience and Command* (New York: Vintage Books, 1971), p. 40.

3. Department of Tactics, *Operations and Training Plan, New Cadet Barracks, 1973* (West Point, New York: U.S. Military Academy, 1973).

4. R. L. Walker, *China Under Communism* (London: Allen and Unwin, 1956), cited in W. Sargant, *Battle for the Mind* (Revised ed., London: Pan Books, 1959), p. 148.

5. Bourne, *op. cit.,* pp. 139–153.

6. J. W. Houston, *Results of First Class Questionnaire, Class of 1971* (West Point, New York: U.S. Military Academy, Office of Institutional Research, 1971), p. 30.

Chapter Four: A Cadet's Four Years

1. *The Fourth-Class System, 1971–1972* (West Point, New York: U.S. Military Academy, 1971).

Chapter Five: Cadet Life

1. D. Cantlay, "Vantage Point," *The Pointer,* December 1971, p. 26.
2. J. P. Sterba, "Dropouts Plague the Air Academy," *The New York Times,* June 3, 1973, p. 1.
3. J. W. Masland and L. I. Radway, *Soldiers and Scholars* (Princeton, New Jersey: Princeton University Press, 1957), p. 201.
4. G. White, "West Point Dismisses Coach Cahill," *The New York Times,* December 14, 1973, p. 61.
5. H. G. Abbott, *Peripheral Neuropathy Technical Report* (West Point, New York: U.S. Military Academy, Medical Research Project, U.S. Army Hospital, 1970).

Chapter Six: Cadets Who Stay and Cadets Who Resign

1. "Dropouts: Farwell to Arms," *Newsweek* (January 21, 1974), p. 98; D. L. Farnsworth, "College Mental Health and Social Change," *Annals of Internal Medicine,* 73: 467–473 (1970); G. W. Medsger, *A Survey of Ex-Cadets of the Class of 1971* (West Point, New York: U.S. Military Academy, Office of Institutional Research, 1971), p. 3.
2. G. W. Medsger, *A Typology for Cadet Leavers* (West Point, New York: U.S. Military Academy, Office of Institutional Research, 1971), pp. 3–4, 5, 12.
3. Medsger, *A Survey of Ex-Cadets of the Class of 1971,* p. 8.
4. D. R. Ilgen, W. Seely, and R. Eggert, *Expectations and NCB Resignations* (West Point, New York: U.S. Military Academy, 1971).

Chapter Seven: A West Point Education

1. S. Huntington, *The Soldier and the State* (New York: Vintage Books, 1957), pp. 198–199.
2. P. Karsten, " 'Professional' and 'Citizen' Officers: A Comparison of Service Academy and R.O.T.C. Officer Candidates," in Moskos, *op. cit.,* pp. 37–57.
3. S. Ambrose, *Duty, Honor, Country* (Baltimore: The Johns Hopkins Press, 1966), p. viii.
4. D. Boroff, "Air Force Academy: A Slight Gain in Altitude" *Harper's Magazine,* February 1963, p. 90.
5. J. Bunting, "Vantage Point." *The Pointer,* November 1971, p. 28.
6. D. Boroff, "West Point: Ancient Incubator for a New Breed" *Harper's Magazine,* December 1962, p. 56.

Chapter Eight: Psychiatry at West Point

1. H. N. Kerns, "Cadet Problems," *Mental Hygiene,* 7:688–696 (1923).
2. *Ibid.*

3. W. H. Friedman and F. W. Coons, "The Mental Health Unit of a Student Health Service: A Study of a Clinic," *Journal of the American College Health Association,* 17:270–283 (1969).

Chapter Nine: Ethics and Honor

1. L. Truscott, "West Point: A Question of Honor," *The New York Times,* August 19, 1972, p. 23.
2. *Ibid.*

Chapter Eleven: Changes

1. B. Weintraub, "British Martial Spirit at Sandhurst Isn't Quite What It Used to Be," *The New York Times,* December 5, 1970, p. 10.
2. Radway, *op. cit.,* p. 13.
3. J. Morgovsky, *The U.S. Military Academy and the Issue of Race in Higher Education: A Perspective* (West Point, New York: U.S. Military Academy, Office of Institutional Research, 1970), p. 9.
4. J. Morgovsky, *Survey of Negro Graduates (One Hundred Years of Blacks Among the Grey)* (West Point, New York: U.S. Military Academy, Office of Institutional Research, 1971).
5. J. Houston, *Some Attitudes of Selected Subsamples of Cadets, Class of 1974* (West Point, New York: U.S. Military Academy, Office of Institutional Research, 1971), pp. 2–7.
6. Morgovsky, *Survey of Negro Graduates, op. cit.*
7. J. Feron, "West Point Simplifies Cadet Regulations," *The New York Times,* August 29, 1973, p. 1.
8. *A Study of the Programs of the United States Military Academy* (West Point, New York: U.S. Military Academy, 1972), p. 36.
9. L. Greenhouse, "5 Accused West Point Cadets Contest Academy Panel's Administration of Honor Code as Unconstitutional," *The New York Times,* May 27, 1973, p. 25; L. Greenhouse, "Honor Code Suit Lost by 6 Cadets," *The New York Times,* June 16, 1973, p. 10.
10. J. Klemesrud, "Kings Point Academy Gets 2 Midshipperson Nominees," *The New York Times,* February 23, 1974, p. 33.
11. J. R. Moskin, "Who Would Ever Go to West Point Today?" *Look,* October 6, 1970, p. 36.

Chapter Twelve: Conclusions

1. W. Just, *Military Men* (New York: Avon Books, 1972), p. 117.
2. M. Janowitz, *The Professional Soldier* (New York: The Free Press, 1960), p. 135.

3. B. Catton, *U.S. Grant and the American Military Tradition* (Boston: Little, Brown and Company, 1954), p. 17.

4. W. E. Woodward, *Meet General Grant* (Garden City, New York: Garden City Publishing Company, 1928), p. 49.

5. *Ibid.,* p. 54.

6. B. H. Liddell Hart, *Sherman: Soldier, Realist, American* (New York: Frederick A. Praeger, 1958), p. viii.

7. R. O'Connor, *Black Jack Pershing* (Garden City, New York: Doubleday and Company, 1961), p. 43.

8. D. C. James, *The Years of MacArthur,* Vol. I (Boston: Houghton Mifflin Company, 1970), p. 72.

9. M. Childs, *Eisenhower: Captive Hero* (New York: Harcourt, Brace and Company, 1959), p. 31.

10. B. H. Liddell Hart, *Why Don't We Learn from History?* (New York: Hawthorn Books, 1971), p. 71.

11. Liddell Hart, *op. cit.,* p. 6.

12. A. Lewis, " 'No Intent to Deceive', " *The New York Times,* July 23, 1973, p. 31.

13. S. M. Hersh, "Military Dispute: Ex-Secretary Rejects Pentagon Version of Cambodian Issue," *The New York Times,* July 19, 1973, p. 1; T. Wicker, "The Big Lie Requires Big Liars," *The New York Times,* July 24, 1973, p. 33.

14. B. D. Ayres, "Army Is Shaken by Crisis in Morale and Discipline," *The New York Times,* September 5, 1971, p. 1; unpublished data, West Point, New York: Office of Institutional Research, 1972.

15. C. F. Bridges, *The Image of the United States Military Academy Among Cadets* (West Point, New York: U.S. Military Academy, Office of Institutional Research, 1972), p. 8.

16. S. M. Hersh, "33 Teachers at West Point Leave Army in 18 Months," *The New York Times,* June 25, 1972, p. 1.

17. Liddell Hart, *Why Don't We Learn From History? op. cit.,* p. 29.

18. J. A. Heise, *The Brass Factories* (Washington, D.C.: Public Affairs Press, 1969).

19. L. Richards, "Oregon Midshipmen Like Naval Academy," *The Sunday Oregonian,* March 17, 1974, p. 22.

Wayne Aldridge

RICHARD C. U'REN graduated from the University of Oregon, then obtained his medical degree from McGill University. He spent two years as an internal medicine resident at the University of Oregon Medical School in Portland, and three years as a resident in psychiatry at Stanford University. In 1970 he went to West Point as chief of the psychiatry service and also held the rank of major in the medical corps. Since September 1972 Dr. U'Ren has been director of the psychiatry clinic and assistant professor of psychiatry at the University of Oregon Medical School. He lives with his wife and three children in Portland.

BIBLIOGRAPHY

Abbott, Howard G. *Peripheral Neuropathy—Technical Report.* West Point, New York: U.S. Military Academy, Medical Research Project, U.S. Army Hospital, 1970.

Ambrose, Stephen E. *Duty, Honor, Country: A History of West Point.* Baltimore: The Johns Hopkins University Press, 1966.

Analysis of the Characteristics of the Class of 1971. West Point, New York: U.S. Military Academy, Office of Institutional Research, 1968.

Bridges, Claude F. *Changes in Commitment of USMA Cadets to a Military Career.* West Point, New York: U.S. Military Academy, Office of Institutional Research, 1969.

Bridges, Claude F. *The Image of the United States Military Academy Among Cadets.* West Point, New York: U.S. Military Academy, Office of Institutional Research, 1972.

Bugle Notes. West Point, New York: U.S. Military Academy, 1971.

Butler, Richard P. *Comparison of USMA Graduates from the Class of 1950 With Graduates from Other College on Selected Variables.* West Point, New York: U.S. Military Academy, Office of Institutional Research, 1971.

Butler, Richard P. *Survey of Careerists and Non-Careerists from the USMA Classes of 1963 Through 1967.* West Point, New York: U.S. Military Academy, Office of Institutional Research, 1971.

Butler, Richard P. and McLaughlin, Gerald W. *Dimensions of Job Satisfaction-Dissatisfaction and Their Relationships to Military Commitment and Retention Among U.S. Military Academy Graduates.* West Point, New York: U.S. Military Academy, Office of Institutional Research, 1971.

Catalogue, 1970/1971. West Point, New York: U.S. Military Academy, 1970.

Catalog, 1973/1974. West Point, New York: U.S. Military Academy, 1973.

Catton, Bruce. *U.S. Grant and the American Military Tradition.* Boston: Little, Brown and Company, 1954.

The Challenge. West Point, New York: U.S. Military Academy, Office of Military Psychology and Leadership, 1971.

Childs, Marquis. *Eisenhower: Captive Hero.* New York: Harcourt, Brace and Company, 1959.

Davis, Kenneth S. *Soldier of Democracy: A Biography of Dwight Eisenhower.* Garden City, New York: Doubleday and Company, 1945.

Donovan, James A. *Militarism, U.S.A.* New York: Charles Scribner's Sons, 1970.

Eisenhower, Dwight D. *In Review.* Garden City, New York: Doubleday and Company, 1969.

Finn, James, ed. *Conscience and Command: Justice and Discipline in the Military.* New York: Vintage Books, 1971.

Forester, C. S. *The General.* Boston: Little, Brown and Company, 1936.

The Fourth-Class System, 1970–1972. West Point, New York: U.S. Military Academy, United States Corps of Cadets, 1971.

Frank, Jerome D. *Persuasion and Healing: A Comparative Study of Psychotherapy.* Baltimore: The Johns Hopkins University Press. Revised edition, 1973.

Friedenberg, Edgar Z. *The Vanishing Adolescent.* New York: Dell Publishing Company, 1962.

Galloway, K. Bruce, and Johnson, Robert Bowie, Jr. *West Point: America's Power Fraternity.* New York: Simon and Schuster, 1973.

Goffman, Erving. *Asylums.* Garden City, New York: Anchor Books, 1961.

Grant, U. S. *Personal Memoirs of U. S. Grant.* Long, E. B., ed. Cleveland and New York: The World Publishing Company, 1952.

Gray, J. Glenn. *The Warriors: Reflections on Men in Battle.* New York: Harper and Row, 1967.

Grinker, Roy R., and Spiegel, John P. *Men Under Stress.* Philadelphia: Blakiston, 1945.

Group for the Advancement of Psychiatry. *Normal Adolescence: Its Dynamics and Impact.* New York: Charles Scribner's Sons, 1968.

Gunther, John. *Eisenhower: The Man and the Symbol.* New York: Harper and Brothers, 1951.

Halberstam, David. *The Best and the Brightest.* New York: Random House, 1972.

Hatch, Alden. *General Ike: A Biography of Dwight D. Eisenhower.* New York: Henry Holt and Company, 1944.

Hausman, William; Sandeson, Richard L.; Wiest, Bernard J.; and Keim, Robert R., Jr. *Adaptation to West Point: A Study of Some Psychological Factors Associated with Adjustment at the United States Military Academy.* West Point, New York: U.S. Military Academy, Military Research Project, U.S. Army Hospital, 1959.

Heise, J. Arthur. *The Brass Factories: A Frank Appraisal of West Point, Annapolis, and the Air Force Academy.* Washington, D.C.: Public Affairs Press, 1969.

Hersh, Seymour. *Cover-up.* New York: Random House, 1972.

Hersh, Seymour. *My Lai 4: A Report on the Massacre and Its Aftermath.* New York: Vintage Books, 1970.

Houston, John W. *Background and Predicted Success of Cadets in the Class of 1973 with Comparisons with Previous Classes.* West Point, New York: U.S. Military Academy, Office of Institutional Research, 1970.

Houston, John W. *Comparison of the USMA Class of 1974 with National Norms for College-Bound Men from Data on the American College Testing Program.* West Point, New York: U.S. Military Academy, Office of Institutional Research, 1970.

Houston, John W. *Report of 1970 Reorganization Week Questionnaire, Classes of 1971, 1972, and 1973.* West Point, New York: U.S. Military Academy, Office of Institutional Research, 1970.

Houston, John W. *Results of First Class Questionnaire, Class of 1971.* West Point, New York: U.S. Military Academy, Office of Institutional Research, 1971.

Houston, John W. *Some Attitudes of Selected Sub-samples of Cadets, Class of 1974.* West Point, New York: U.S. Military Academy, Office of Institutional Research, 1971.

Houston, John W. *Trends in Admission Variables Through the Class of 1975.* West Point, New York: U.S. Military Academy, Office of Institutional Research, 1971.

Houston, John W., and Cooke, James L. *Characteristics of the Class of 1974.* West Point, New York: U.S. Military Academy, Office of Institutional Research, 1970.

Houston, John W.; Fabian, John M.; and Greco, David L. *Characteristics of the Class of 1975.* West Point, New York: U.S. Military Academy, Office of Institutional Research, 1971.

Houston, John W., and Greco, David L. *Report of New Cadet Barracks Questionnaire, Class of 1974.* West Point, New York: U.S. Military Academy, Office of Institutional Research, 1971.

Houston, John W.; Stoller, Daniel L.; and Hespenheide, John E. *Characteristics of the Class of 1973.* West Point, New York: U.S. Military Academy, Office of Institutional Research, 1969.

Huntington, Samuel P. *The Soldier and the State: The Theory and Politics of Civil-Military Relations.* New York: Vintage Books, 1957.

Hutchins, Robert Maynard. *The Higher Learning in America.* New Haven: Yale University Press, 1962.

Ilgen, Daniel R.; Seely, William; and Eggert, Richard. *Expectations and NCB Resignations.* West Point, New York: U.S. Military Academy, 1971.

James, D. Clayton. *The Years of MacArthur. Volume I: 1880–1941.* Boston: Houghton Mifflin Company, 1970.

Janowitz, Morris. *The Professional Soldier.* New York: The Free Press, 1960.

Janowitz, Morris, ed. *The New Military: Changing Patterns of Organization.* New York: Russell Sage Foundation, 1964.

Janowitz, Morris, and Little, Roger, eds. *Sociology and the Military Establishment.* New York: Russell Sage Foundation, revised edition, 1965.

Johnson, Haynes, and Wilson, George C. *Army in Anguish.* New York: Pocket Books, 1972.

Just, Ward. *Military Men.* New York: Avon Books, 1972.

King, Charles. *The True Ulysses S. Grant.* Philadelphia: J. B. Lippincott Company, 1914.

King, Edward L. *The Death of the Army: A Pre-Mortem.* New York: Saturday Review Press, 1972.

Kolb, Lawrence C. *Modern Clinical Psychiatry.* Philadelphia: W. B. Saunders Company, eighth edition, 1973.

Liddell Hart, B. H. *Sherman: Soldier, Realist, American.* New York: Fredrick A. Praeger, 1958.

Liddell Hart, B. H. *Why Don't We Learn From History?* New York: Hawthorn Books, 1971.

Lifton, Robert Jay. *Death in Life: Survivors of Hiroshima.* New York: Vintage Books, 1969.

MacArthur, Douglas. *Reminiscences.* New York: McGraw-Hill, 1964.

McCormick, Robert R. *Ulysses S. Grant.* New York: D. Appleton-Century Company, 1934.

McLaughlin, Gerald W., Jr. *A Multidimensional View of Cadets' Decision to Seek a USMA Nomination.* West Point, New York: U.S. Military Academy, Office of Institutional Research, 1970.

McLaughlin, Gerald W., Jr., and Butler, Richard P. *Perceived Importance of Various Job Characteristics by West Point Graduates.* West Point, New York: U.S. Military Academy, Office of Institutional Research, 1971.

Marron, Joseph E. *The Frequency, Source, and Value of Cadet Surveys (A Research Note).* West Point, New York: U.S. Military Academy, Office of Institutional Research, 1971.

Marron, Joseph E. *Some Correlates of the Aptitude for the Service Rating (A Research Note).* West Point, New York: U.S. Military Academy, Office of Institutional Research, 1971.

Masland, John W., and Radway, Laurence I. *Soldiers and Scholars: Military Education and National Policy*. Princeton, New Jersey: Princeton University Press, 1957.

Medsger, Gerald W. *A Survey of the Ex-Cadets of the Class of 1971*. West Point, New York: U.S. Military Academy, Office of Institutional Research, 1971.

Medsger, Gerald W. *A Typology for Cadet Leavers*. West Point, New York: U.S. Military Academy, Office of Institutional Research, 1971.

Merrill, James M. *William Tecumseh Sherman*. New York: Rand McNally and Company, 1971.

Military Psychiatry. (Technical Manual 8-244.) Washington, D.C.: Department of the Army, 1957.

Morgovsky, Joel. *Survey of Negro Graduates (One Hundred Years of Blacks Among the Grey)*. West Point, New York: U.S. Military Academy, Office of Institutional Research, 1971.

Morgovsky, Joel. *Trends in Responses to First Class Questionnaire, 1969, 1970, 1971 (A Research Note)*. West Point, New York: U.S. Military Academy, Office of Institutional Research, 1971.

Morgovsky, Joel. *The U.S. Military Academy and the Issue of Race in Higher Education: A Perspective*. West Point, New York: U.S. Military Academy, Office of Institutional Research, 1970.

Morgovsky, Joel. *What Do Parents Know About West Point?* West Point, New York: U.S. Military Academy, Office of Institutional Research, 1970.

Moskos, Charles C., Jr., ed. *Public Opinion and the Military Establishment*. Beverly Hills, California: Sage Publications, 1971.

Mumford, Lewis. *The Condition of Man*. London: Martin Secker and Warburg Ltd., 1944.

O'Connor, Richard. *Black Jack Pershing*. Garden City, New York: Doubleday and Company, 1961.

Operations and Training Plan, New Cadet Barracks, 1973. West Point, New York: U.S. Military Academy, Departments of Tactics, 1973.

Oppenheimer, Martin, ed. *The American Military*. New York: Transaction Books, 1971.

Palmer, Frederick. *John J. Pershing*. Harrisburg, Pennsylvania: The Military Service Publishing Company, 1948.

Register of Graduates and Former Cadets of the United States Military Academy. West Point, New York: The West Point Alumni Foundation, Inc., 1970.

Regulations for the United States Corps of Cadets. West Point, New York: U.S. Military Academy, Commandant of Cadets, 1971.

Regulations for the United States Corps of Cadets. West Point, New York: U.S. Military Academy, Commandant of Cadets, 1973.

Remarque, Erich Maria. *All Quiet on the Western Front*. New York: Grosset and Dunlap, 1929.

Sanford, Nevitt, ed. *The American College: A Psychological and Social Interpretation of the Higher Learning.* New York: John Wiley and Sons, 1962.

Sanford, Nevitt. *Self and Society: Social Change and Individual Development.* New York: Atherton Press, 1966.

Sargant, William. *Battle for the Mind.* London: Pan Books, 1959.

Sargant, William. *The Unquiet Mind.* London: Pan Books, 1971.

Spenser, Gary. *A Social-Psychological Profile of the Class of 1973—A First Report.* West Point, New York: U.S. Military Academy, Office of Institutional Research, 1969.

A Study of the Programs of the U.S. Military Academy. West Point, New York: U.S. Military Academy, 1972.

Tiger, Lionel. *Men in Groups.* New York: Random House, 1969.

Tiger, Lionel, and Fox, Robin. *The Imperial Animal.* New York: Holt, Rinehart and Winston, 1971.

Von Clausewitz, Carl. *On War.* Anatol Rapoport, ed. Baltimore: Penguin Books, 1968.

Wise, Arthur E. *A Comparison of New Cadets at USMA with Entering Freshmen at Other Colleges.* West Point, New York: U.S. Military Academy, Office of Institutional Research, 1969.

Woodward, W. E. *Meet General Grant.* Garden City, New York: Garden City Publishing Company, 1928.

Yarmolinsky, Adam. *The Military Establishment: Its Impacts on American Society.* New York: Harper and Row, abridged edition, 1973.

INDEX